Optical Crystallography

By ERNEST E. WAHLSTROM

Introduction to Theoretical
Igneous Petrology

Igneous Minerals and Rocks

Optical Crystallography
Second Edition

Optical
Crystallography

ERNEST E. WAHLSTROM
Department of Geology and Mineralogy
University of Colorado

Second Edition

New York · JOHN WILEY & SONS, Inc.
London · CHAPMAN & HALL, Ltd.

COPYRIGHT, 1943, 1951, BY
ERNEST E. WAHLSTROM

All Rights Reserved

This book or any part thereof must not be reproduced in any form without the written permission of the publisher.

PRINTED IN THE UNITED STATES OF AMERICA

Preface

Optical crystallography is a rapidly growing field. The study of crystal optics is no longer confined to the mineralogist and the petrographer. Although the fundamental theories were developed by early optical crystallographers who perfected polarizing microscopes for the examination of thin slices of rocks, scientists in many other fields of investigation have come to recognize the value of these powerful tools for research.

Optical crystallography contributes greatly to research in many branches of chemistry. It is used chiefly in the rapid identification of solids, whether organic or inorganic. Metallurgists, ceramists, and workers in related fields accept optical theory and instruments as indispensable parts of their working equipment. The medical profession makes use of the polarizing microscope in its battle against industrial diseases such as silicosis. Engineers obtain critical data from the microscopic examination of many structural materials with which they work.

It is the purpose of this textbook to review the principles of optical crystallographic theory. Practical applications are not given a prominent place because I feel that, once a student has firmly grasped fundamental concepts, he will be able to meet any problem related to the field. Some space is given to a description of the techniques for the measurement of refractive indices. Emphasis is placed on the immersion method of index measurement, for this technique is the most widely used.

Since the first edition of *Optical Crystallography* appeared, in 1943, I have gained the valuable experience of using this textbook with large numbers of students of varied backgrounds and aptitudes. During this period, I have also received many comments and criticisms from colleagues and professional workers who, for one reason or another, have examined the book critically and in detail.

The reader will observe that the text is profusely illustrated. Three-dimensional visualization is required of the student who wishes to master the theory of optical crystallography. With this in mind, I have abandoned many of the two-dimensional illustrations in favor of block diagrams. All diagrams in which the third dimension has been indicated were drawn in clinographic projection; this type of projection is familiar to the crystallographer but may present a

v

slightly distorted appearance to one who is accustomed to block diagrams drawn in perspective.

No effort is made to include descriptive tables of minerals or crystalline chemicals. Optical descriptions of most natural and artificial compounds are to be found in many standard references on mineralogy and chemistry.

I have gathered the material for this book from many sources, of which the more important are listed in Selected References. The illustrations are for the most part original, but, of course, many of the diagrams were suggested by illustrations appearing in other published works in the field of optical crystallography.

I wish to extend thanks to the numerous individuals who have offered criticisms and suggestions. Many of the improvements in this edition are the result of friendly collaboration with students and professional colleagues. It is with a deep feeling of obligation that I acknowledge the indispensable assistance given to me by my co-worker, Kathryn Kemp Wahlstrom, who patiently worked at my side in all stages of preparation and editing of the manuscript.

ERNEST E. WAHLSTROM

Boulder, Colorado
April, 1951

Contents

Chapter I

CRYSTALLOGRAPHY

Nature of Crystals. A *crystal* may be defined as a polyhedral solid bounded by plane faces which express an orderly internal arrangement of atoms or molecules. In the study of the internal structure of substances by X-ray techniques, less emphasis is placed on crystal faces; a *crystal* is regarded as a body characterized by a more or less undisturbed three-dimensional space extension of a characteristic unit of internal structure. A distinction is made between *crystalline substances* and *amorphous substances.* Amorphous substances display random arrangement of atoms or molecules.

If emphasis is placed on the presence or absence of crystal faces, the following distinctions apply: *euhedral* crystals possess a completely developed array of faces; *subhedral* crystals show a partial development of faces; no crystal faces are present on *anhedral* grains.

Law of Constancy of Interfacial Angles. The vast majority of crystals are malformed. Growth conditions cause crystals to develop unsymmetrically. Crystals precipitated simultaneously from the same solution rarely look exactly alike. *However, in a given chemical or mineral species, no matter what growth irregularities are present, the angles between similarly chosen adjacent or projected faces are constant.*

Crystal Axes and Crystal Systems. Crystal faces are conveniently referred to imaginary lines or directions which may be used to describe the position of a face or group of faces in space. These lines or directions are called *crystal axes* (Fig. 1). All crystals naturally fall into six systems, based on six simple geometric groupings of the crystal axes. The six systems are as follows:

I. *Isometric system.* Crystals in this system are referred to three mutually perpendicular, equal axes. The axes are designated as a_1, a_2, and a_3.

II. *Tetragonal system.* Crystals referable to three mutually perpendicular axes, two equal and one either longer or shorter, belong in this system. The axes are designated as a_1, a_2, and c.

III. *Hexagonal system.* This system includes what some crystallographers call the trigonal system. It embraces all crystals which

are referred to four axes: three of these lie in a plane, intersect at 60- and 120-degree angles, and are equal in length; the fourth axis is perpendicular to the plane including the other three and is either longer or shorter than the other axes. The axis designation is a_1, a_2, a_3, and c.

IV. *Orthorhombic system.* This system includes all crystals referable to three unequal, mutually perpendicular axes. The axes are designated as a, b, and c. By convention, orthorhombic crystals are

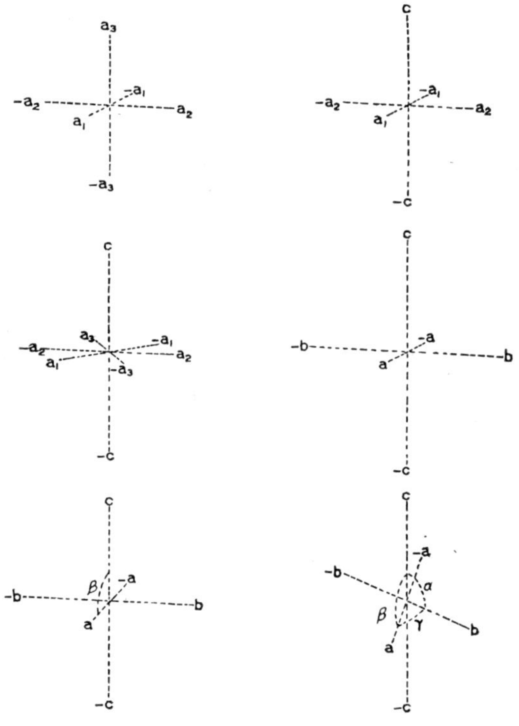

FIG. 1. Axes of six crystal systems.

oriented so that the unit intercept on the a axis, called the *brachy* axis, is shorter than the unit intercept on the b axis, called the *macro* axis.

V. *Monoclinic system.* This system contains crystals referred to three unequal axes: two are in a plane and intersect at acute and obtuse angles; the third axis is perpendicular to the plane including the other two. The a and b axes are called the *clino* and *ortho* axes, respectively. The obtuse angle between the positive ends of the a and c axes is identified as β.

VI. *Triclinic system.* This system contains all crystals that cannot be placed in the above systems. Its three unequal axes intersect at acute and obtuse angles. The axes are designated as a, b, c; as in the orthorhombic system, the a axis is called the *brachy* axis, and the b axis the *macro* axis. The angles between the positive ends of b and c, c and a, and a and b are designated as α, β, and γ, respectively.

Axial Ratio. The axial ratio of a mineral or crystalline chemical is characteristic for each species. Commonly the axial ratio is determined by choosing a prominent crystal face that cuts all three axes and then calculating the relative intercepts on the axes. The face chosen for this calculation, if it is a pyramid face, is called the *unit pyramid.* Modern X-ray technique assists in the choice of the proper face for the computation of the axial ratio. As a matter of fact, the axial ratio may be determined in certain substances by the X-ray method without considering the external form of the crystal.

In the isometric system all axes are equal; hence the axial ratio for all isometric crystals is the same. In tetragonal crystals the lateral axes are equal, but the vertical axis is longer or shorter. All that is necessary is a statement of the intercept on the vertical axis relative to the intercept on the lateral axes; for example, $c = 1.1321$ indicates that the intercept on the c axis is to the intercept on a lateral axis as 1.1321 is to 1. The same type of reasoning applies to the hexagonal system. In the orthorhombic system the intercept on the b axis is set equal to unity, and the axial ratio is stated thus: $a{:}b{:}c = 0.8131{:}1{:}1.2034$.

In the monoclinic and triclinic systems, it is necessary to state not only the axial ratio but also the angular relationships of the axes.

Law of Rational Intercepts. Once the unit intercepts have been established, the position of any face on a crystal may be described by ascertaining its intercepts on each of the axes as related to the unit intercepts. In making this evaluation, the *law of rational intercepts* is useful. This law states that *the ratios between the intercepts for crystal faces must be rational numbers,* that is, $1{:}2$, $3{:}\frac{3}{2}$, $4{:}\frac{2}{3}$, etc., but never $1{:}\sqrt{2}$, etc.

Parameters and Indices. The *parameters* of a crystal face express by a series of numbers the relative intercepts by that face on the crystallographic axes. The relative intercepts are expressed in terms of the unit intercepts. For example, the parameters of the unit pyramid in the orthorhombic system are $a{:}b{:}c$. Another pyramid might have the parameters $\frac{1}{3}a{:}\frac{1}{4}b{:}c$.

Miller indices are the reciprocals of the parameters cleared of fractions. The relationships between parameters and indices are indicated in the following examples.

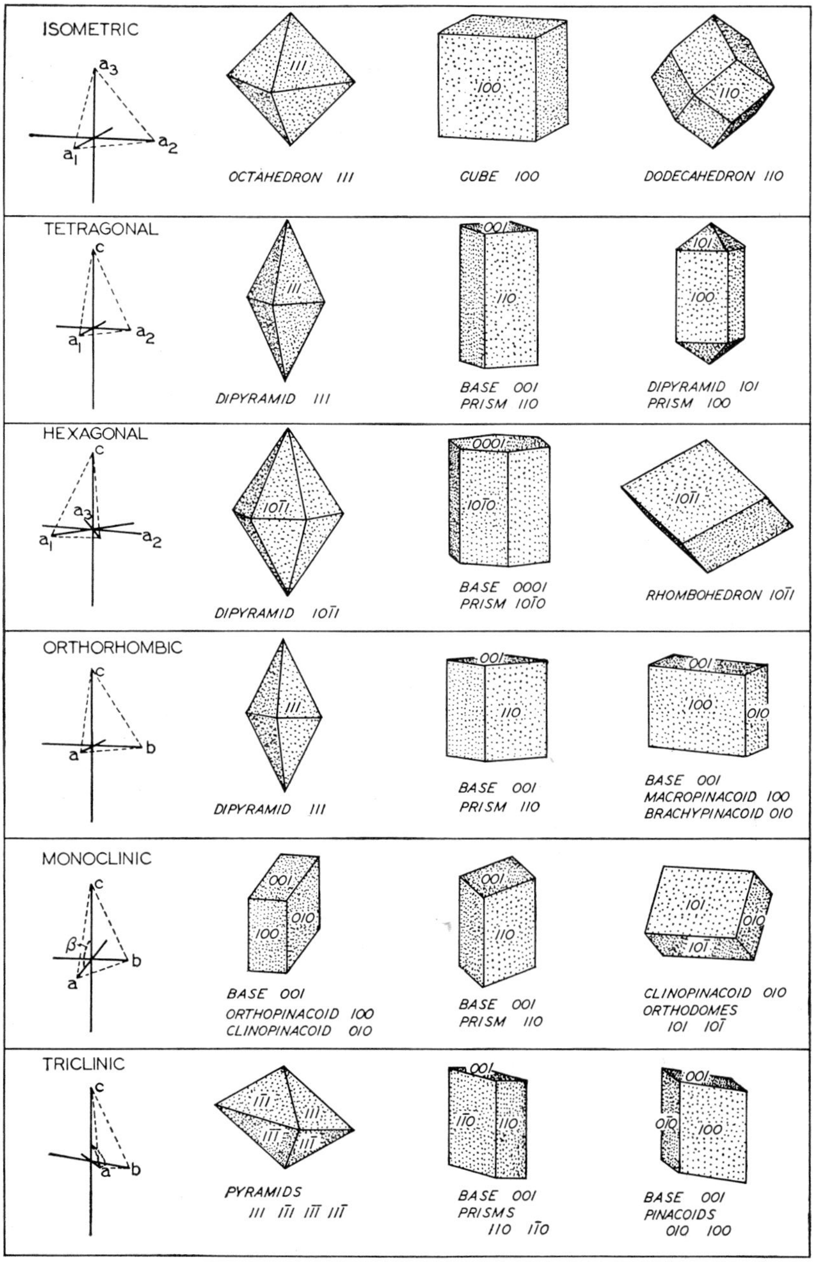

FIG. 2. Axes, simple forms, and combinations in the six crystal systems.

Parameters	*Miller Indices*
$a_1:a_2:a_3$	111
$\infty a_1:\infty a_2:c$	001
$a_1:\infty a_2:-a_3:c$	$10\bar{1}1$
$\frac{1}{2}a:\frac{2}{3}b:c$	432
$\infty a:\frac{3}{5}b:c$	053
$\frac{1}{3}a_1:\frac{1}{3}a_2:c$	331
$a_1:a_2:3c$	331

Crystal faces are more easily represented by indices than by parameters; hence indices are used almost exclusively. The indices, in generalized form, may be designated by the letters h, k, i, and l.

Form. Crystallographically, a *form* is a face or group of faces which bear like relationships to the crystallographic axes. For example, a cube is a form consisting of six similar faces, each of which is perpendicular to one axis of the isometric system and parallel to the other two. When two or more forms are present on a crystal, it is said to be a *combination.*

Figure 2 shows simple forms and combinations in the six crystal systems.

When indices are used to designate individual faces, the indices are enclosed in parentheses. Thus $(1\bar{2}1)$, $(h0l)$, (hkl) designate individual faces. If the indices are enclosed in braces thus, $\{0kl\}$, $\{120\}$, $\{hkl\}$, $\{hk\bar{\imath}l\}$, they refer to a complete form rather than to an individual face of a form.

For convenience, forms as represented in crystal drawings commonly are indicated by letters or standard symbols.

Symmetry Elements. The elements of symmetry include center of symmetry, planes of symmetry, and axes of symmetry. A crystal is said to have a *center of symmetry* if a line passed from any point on the surface of the crystal through the center of the crystal emerges at a similar point on the opposite side at the same distance from the center. A *plane of symmetry* is present if an imaginary plane can be passed through a crystal so as to divide it into symmetrical halves, each the mirror image of the other. A *symmetry axis* is an imaginary line about which a crystal may be rotated so as to bring identical faces, lines, or angles into view at least twice during a complete rotation. The symmetry elements are best visualized in perfectly formed symmetrical crystals (Fig. 3).

There are 32 possible combinations of symmetry elements, which give rise to 32 crystal classes.

Habit. The shape acquired by a crystal depends on many factors, such as the temperature, pressure, and the composition of the parent

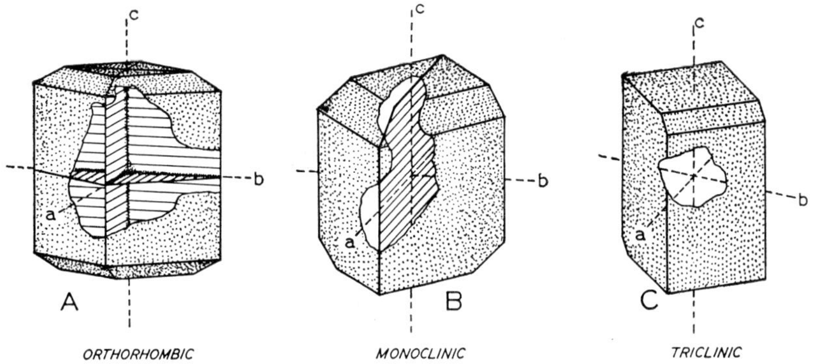

FIG. 3. Elements of symmetry in selected crystals.

A. Orthorhombic crystal showing three planes of symmetry (ruled), three axes of twofold symmetry, each parallel to a crystallographic axis, and a center of symmetry.

B. Monoclinic crystal with one plane of symmetry (ruled), a twofold axis parallel to the *b* axis, and a center of symmetry.

C. Triclinic crystal having a center of symmetry only.

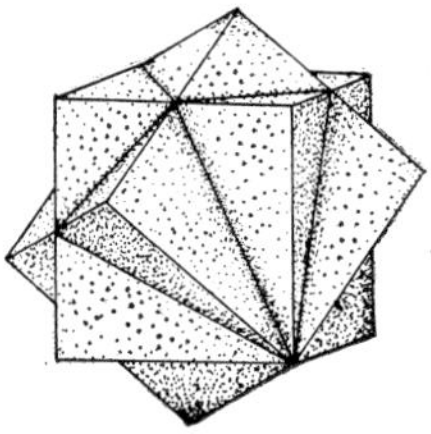

FIG. 4

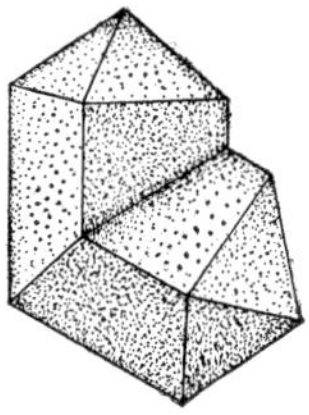

FIG. 5

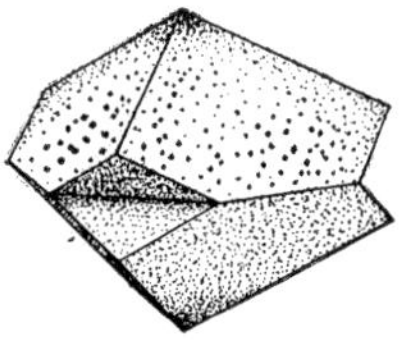

FIG. 6

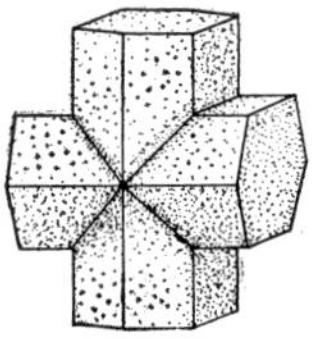

FIG. 7

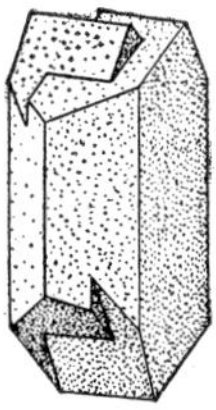

FIG. 8

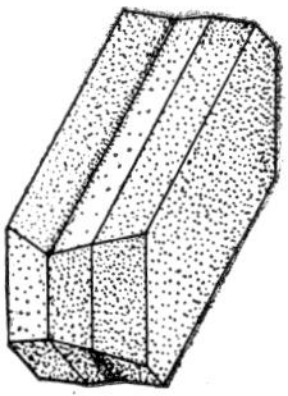

FIG. 9

TYPICAL TWINS

FIG. 4. Penetration twin. Cube twinned on {111}. Isometric.
FIG. 5. Elbow twin on {101}. Tetragonal.
FIG. 6. Rhombohedron twinned on {0001}.
FIG. 7. Penetration twin. Twinned on {032}. Orthorhombic.
FIG. 8. Carlsbad penetration twin of feldspar. Twinned on {100}. Monoclinic.
FIG. 9. Albite trill. Twinned on {010}. Triclinic.

solutions. Moreover, impurities, movements of solvents, differences in concentration from place to place, and the rate of precipitation from solution contribute to the variations observed in crystals. However, precipitation of a given compound generally results in a characteristic shape or outline, or *habit*. Expressions commonly used to denote

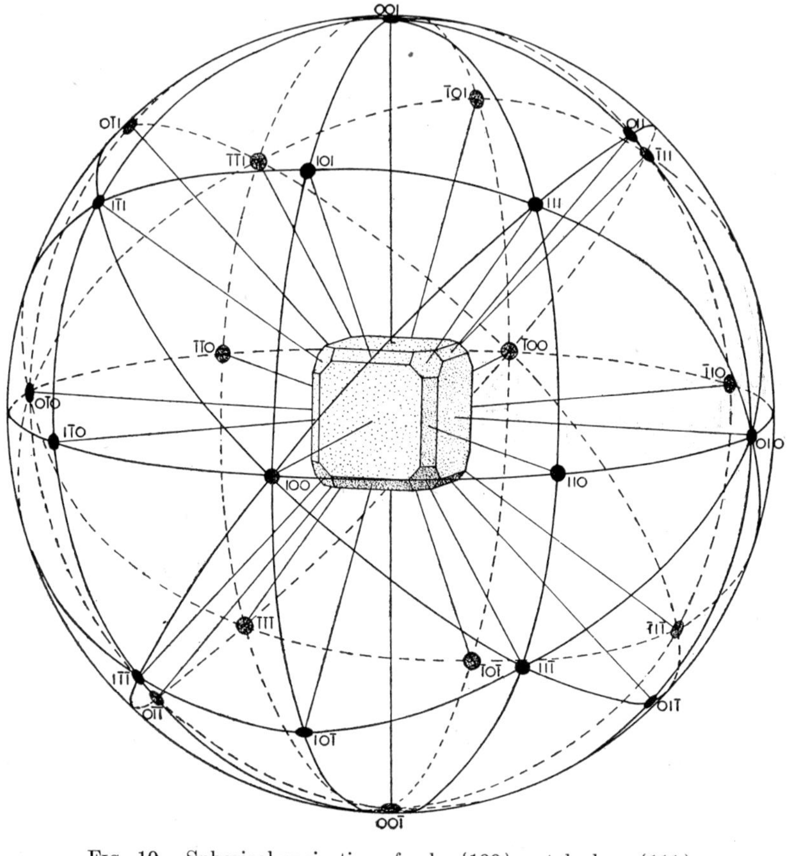

Fig. 10. Spherical projection of cube {100}, octahedron {111}, and dodecahedron {110}.

habit are the following: *tabular, platy, micaceous, equant* or *equidimensional, stubby, prismatic, acicular,* and *fibrous.*

Twins. Twin crystals result from the intergrowth of two or more crystals according to some definite law. One part of a twin crystal is related to another as if it were rotated 180 degrees with respect to the other part about a crystal direction common to both. The axis about which the rotation appears to take place is called the *twin axis;*

the plane involved, perpendicular to the twin axis, is called the *twin plane.* The plane along which the two individuals unite is called the *composition plane.*

Twin crystals may be either *contact* or *penetration twins.* A *simple twin* is twinned once. *Repeated* or *polysynthetic twins* are twinned

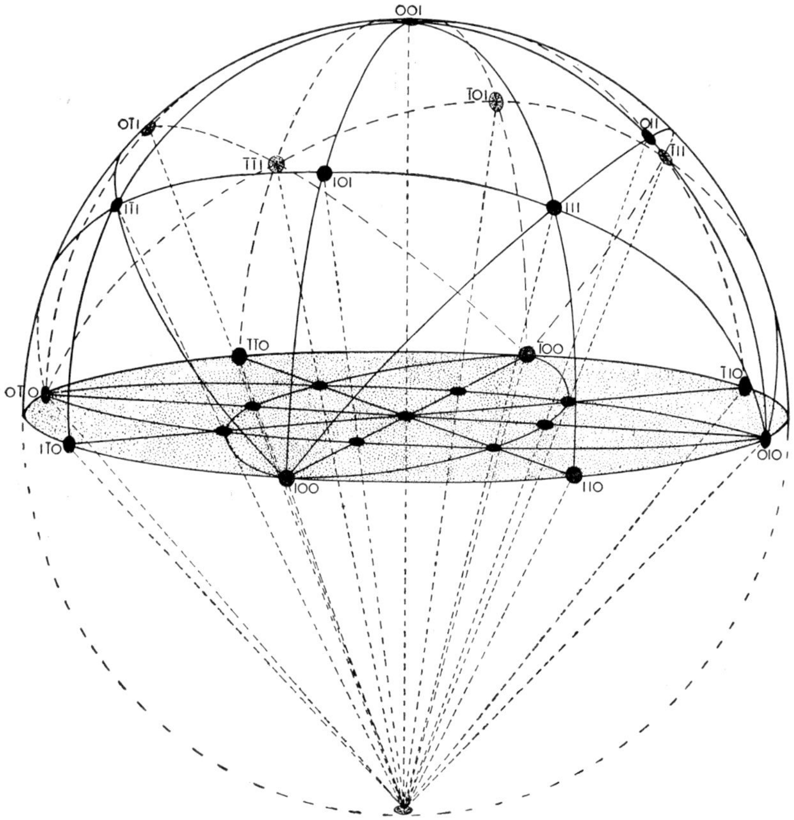

Fig. 11. Diagram showing relationship of spherical projection to stereographic projection (stippled plane).

two or more times and are designated as *trills, fourlings, fivelings,* etc. Figures 4 to 9 show typical twins in each of the six crystal systems.

Projections. Projections play an important part in the graphic depiction of space relationships of crystal faces and directions. Projections have an increasingly important use in optical crystallography. There are various types, each serving best a specific purpose.

Figure 10 shows the *spherical projection* of an isometric crystal—

a combination of cube, octahedron, and dodecahedron. To construct the spherical projection the crystal is enveloped by a sphere whose center is coincident with that of the crystal, and normals to each crystal face are drawn out from the center until they intersect the surface of the sphere. The points of intersection of the normals and

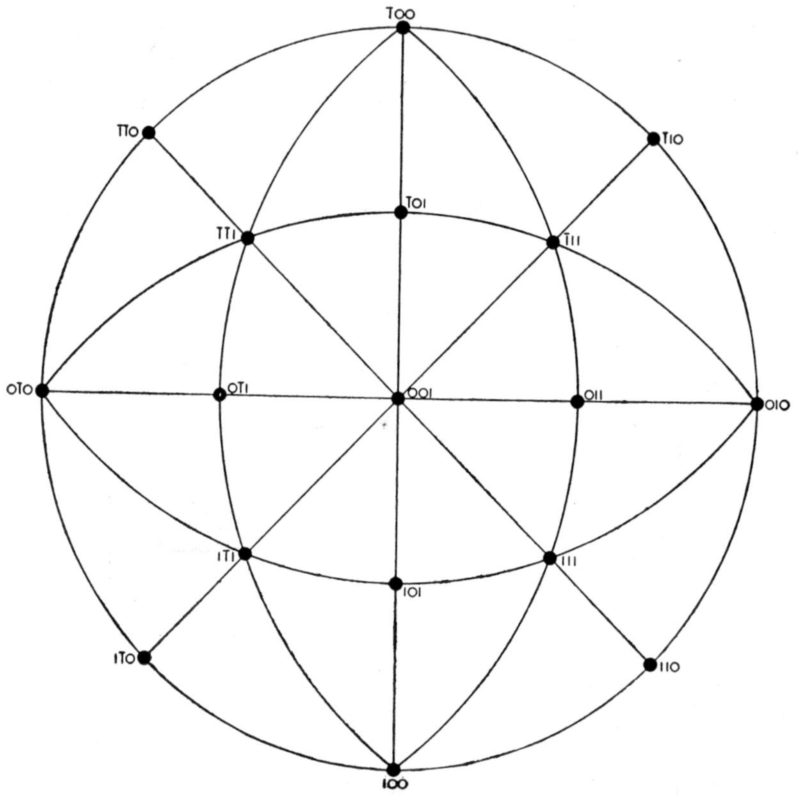

FIG. 12. Stereographic plot of poles of cube {100}, octahedron {111}, and dodecahedron {110}.

the sphere locate the *poles* of the faces. Crystal faces whose edges of intersection are parallel have poles which lie on a great circle of the sphere. Great circles which contain two or more poles are designated as *zones*. The spherical projection is three-dimensional and cannot be used with ease for routine portrayal of crystal faces or directions.

In ordinary work the poles are projected into a plane for two-dimensional representation. The choice of the position of the plane

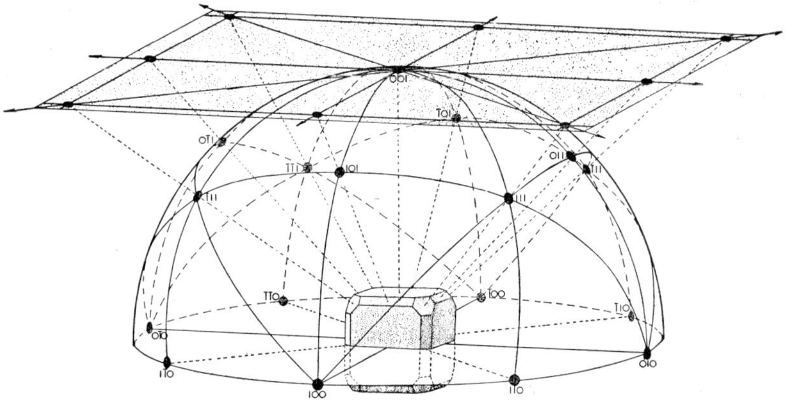

FIG. 13. Diagram showing relationship of spherical projection to gnomonic projection (stippled plane).

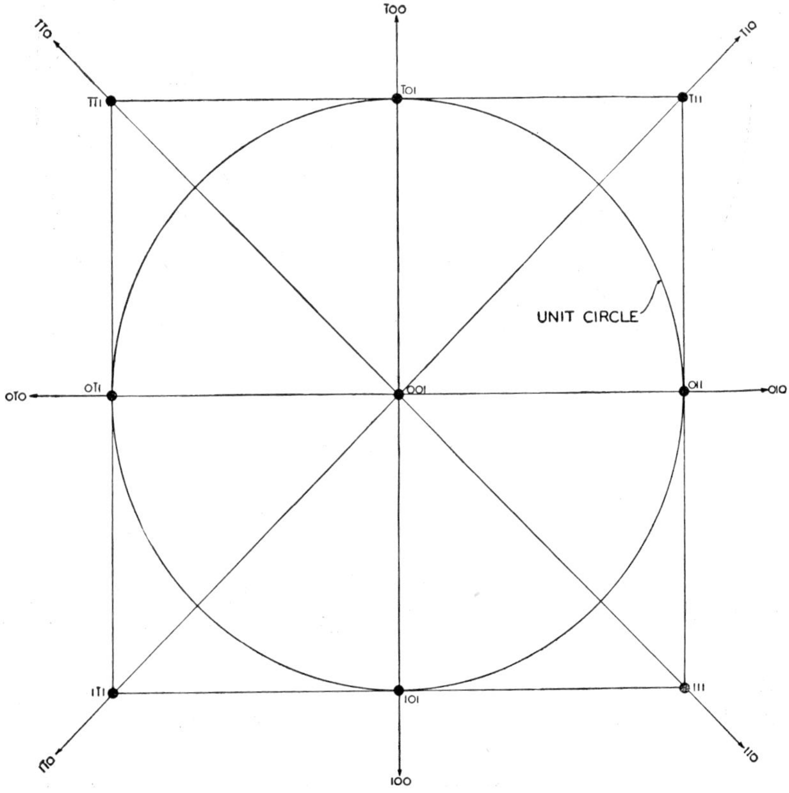

FIG. 14. Gnomonic projection of poles of cube, octahedron, and dodecahedron.

and the manner of projection into it are determined by the use to which the projection is to be put.

One of the most useful projections is the *stereographic projection.* In this type the plane of projection is the equatorial plane of the spherical projection. The location of each pole in the stereographic

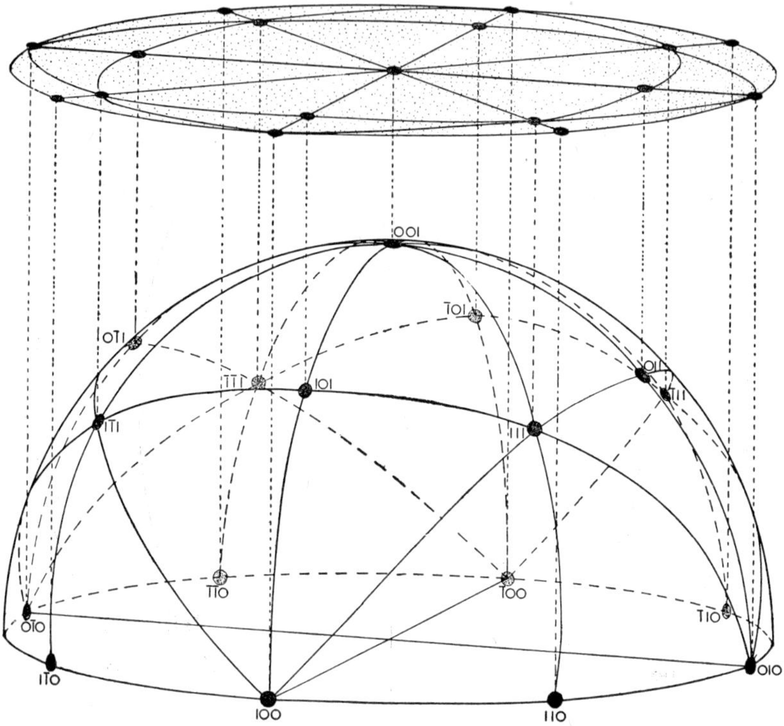

Fig. 15. Diagram showing relationship of spherical projection to orthographic projection (stippled plane).

projection is obtained by determining the location of the point of intersection with the equatorial plane of a line drawn from the pole on the spherical projection to its south pole. The construction of the stereographic projection of a combination of cube, octahedron, and dodecahedron is shown in Fig. 11. Figure 12 shows the stereographic projection in plan.

North-south meridian zones of the spherical projection appear as radial lines in the stereographic projection, and other great circles or arcs of circles are transferred to the stereographic projection as arcs of true circles. The latter property permits easy geometric construction.

Stereographic projections are particularly useful in the representation of the geometric relations between crystal directions and optical properties. For example, in the triclinic system the stereographic projection may be used to indicate the orientation of the optical indicatrix with respect to the crystallographic axes of certain crystal faces.

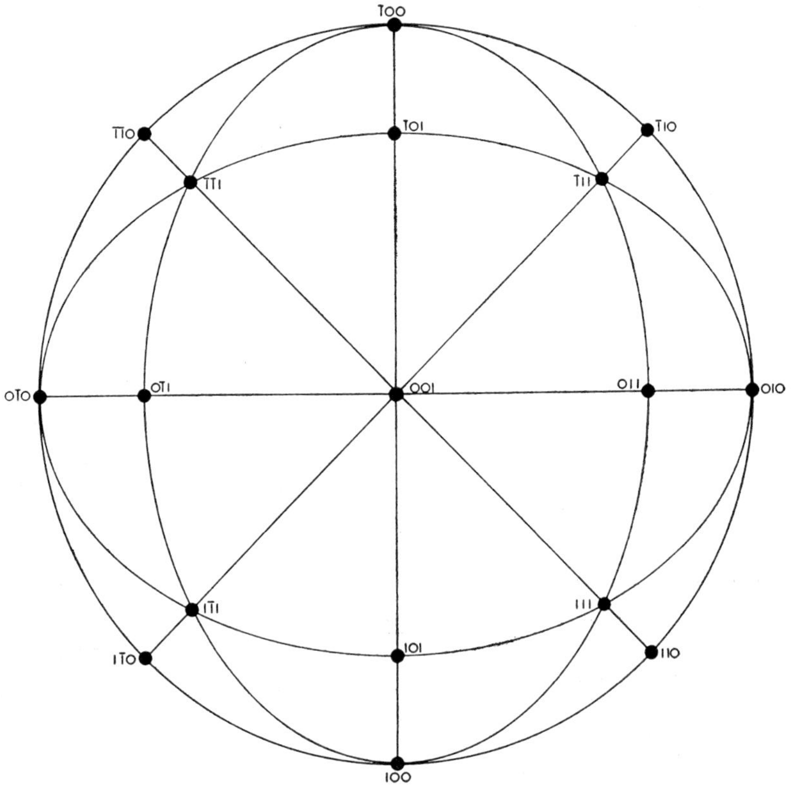

FIG. 16. Orthographic projection of poles of cube, octahedron, and dodecahedron.

Like the stereographic projection, the *gnomonic projection* is derived from the spherical projection. The plane of projection, however, is tangent to the north pole of the sphere. Figure 13 shows the relations of the gnomonic projection to the spherical projection. It will be noted that all zonal lines in the gnomonic projection are straight lines. Poles of faces parallel to the vertical axis of the sphere lie at infinity in the gnomonic projection and are designated by straight lines terminated by arrows. Figure 14 shows the gnomonic projection of a cube, octahedron, and dodecahedron.

The gnomonic projection is widely used to plot data obtained by measurements of crystals with a two-circle goniometer. The projection readily lends itself to computation of axial ratios and crystal drawing.

The *orthographic projection* is obtained by dropping normals from the poles in the spherical projection to the plane of projection as in Fig. 15, in which the plane of projection is normal to the north-south axis of the sphere. However, the plane may be put into any desired position with respect to the spherical projection in order to serve special requirements.

The orthographic projection is constructed with difficulty because, in general, great circles or arcs of great circles on the spherical projection in the orthographic projection become ellipses or arcs of ellipses. Figure 16 shows an orthographic projection of a cube, octahedron, and dodecahedron.

The orthographic projection is particularly useful in the study of the origin of interference phenomena under the polarizing microscope.

CHAPTER II

PHYSICAL PROPERTIES

Introduction. Preliminary determination of the physical properties other than the optical properties aids in the identification of crystals or fragments under a polarizing microscope. Important properties are cleavage, parting, fracture, hardness, specific gravity, color, luster, and streak. Fusibility occasionally may be determined.

Relation of Physical Properties to Crystal Direction. Physical properties reflect crystal structure and symmetry in detail. For example, hardness, or resistance to abrasion, is constant on all similar faces in the isometric system. But, as is shown by diamond, unlike crystal faces have different degrees of hardness. In crystals belonging to systems with unequal crystallographic axes, the hardness is different in each of the crystallographic directions. Cleavage, for example, is dependent upon internal atomic arrangement, which, in turn, controls the external form of a crystal. The nature of light reflected from one face of a crystal may differ from that reflected from another unlike face. Of importance to the optical crystallographer is the fact that the manner of passage of light through a crystal is a guide to the symmetry of the crystal.

Cleavage, Parting, and Fracture. *Cleavage* is the tendency of a crystalline substance to split along one or more crystal directions. Plane surfaces resulting from splitting along cleavage directions ordinarily are parallel to common crystal faces. Cleavage in a crystal species is a constant property which expresses a characteristic internal structure.

A crystal may possess several cleavages or none. The quality of cleavage may range from poor to perfect. The number of cleavages and the quality of each do not vary in a given mineral or chemical species no matter what the external form or occurrence may be. Figures 1 to 4 illustrate several types of cleavage commonly seen in crystalline substances.

Parting is the tendency of certain crystalline substances to split along smooth planes which do not necessarily parallel crystal planes or faces. Parting is not a constant property but is often controlled by regularly oriented inclusions which cause planes of weakness to form.

External causes may induce a tendency to split which was not originally present.

Fracture defines the manner in which a substance breaks in any direction that is not a cleavage direction. Commonly used terms de-

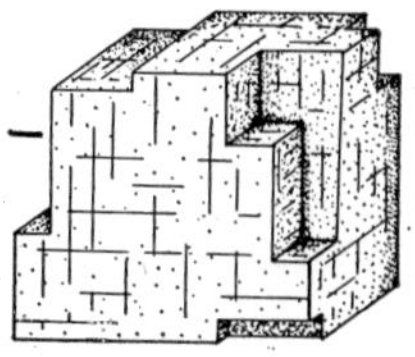

FIG. 1. Cubic cleavage.

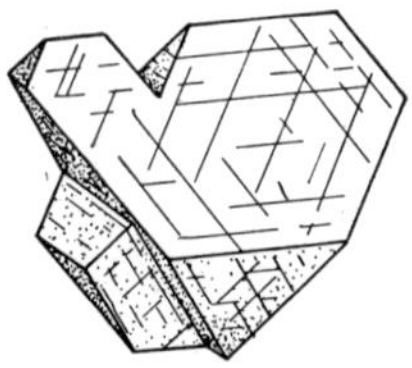

FIG. 2. Octahedral cleavage.

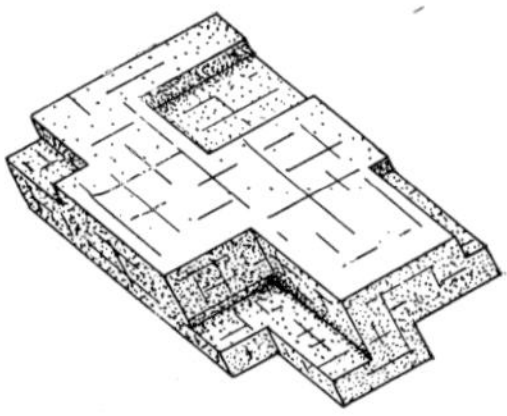

FIG. 3. Rhombohedral cleavage.

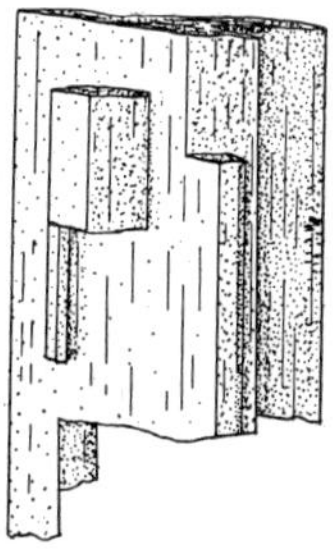

FIG. 4. Prismatic cleavage.

noting fracture are *conchoidal, subconchoidal, uneven, splintery,* and *hackly.*

Hardness. *Hardness* expresses the degree of resistance of a substance to abrasion. The hardness of a crystal reflects the intensity of the forces which bind the atoms and molecules of the crystal together. In general, hard substances are relatively insoluble.

There are many simple and elaborate devices for testing hardness. Many standards have been devised, but none is more useful than the scale of hardness used by mineralogists, the *Mohs scale.*

This scale is based on the following ten minerals:

1. Talc
2. Gypsum
3. Calcite
4. Fluorite
5. Apatite
6. Orthoclase
7. Quartz
8. Topaz
9. Corundum
10. Diamond

Each mineral will scratch any mineral above it in the scale. For example, a mineral with a hardness of 6.5 will scratch orthoclase but

not quartz. Using the Mohs scale these substances have the following hardness:

Fingernail	2.5
Brass pin	3.5
Cover glass	5.5–6
Knife blade	5.5–6
File	6.5

Specific Gravity. *Specific gravity* is the ratio of the weight of a given volume of a substance to the weight of an equal volume of water. The determination of specific gravity rests on the following fundamental equation:

$$\text{Sp gr} = \frac{W}{W - W_1}$$

where W is the weight of a solid in air, and W_1 is the weight of the same solid in water.

Several devices are available for specific-gravity determination. These include the *Jolly balance, beam* or *Westphal balance,* and *pycnometer.* Noteworthy are recently devised microbalances for the determination of specific gravity of very small crystals or fragments. Many of these balances employ some immersion medium other than water.

Color. *Color* depends upon the nature of the light absorbed at the surface or transmitted by a substance. If a substance is white, it reflects or transmits all colors of the spectrum equally. A red mineral, for example, reflects or transmits the red wave length of the spectrum but absorbs or suppresses the other wave lengths. Black substances absorb all wave lengths.

Color is a variable property which in many substances depends on the presence of impurities, on internal structure, or on the state of subdivision.

Luster. *Luster* relates to the nature and quantity of light reflected from the surface of a substance. The following are important lusters of nonmetallic materials: vitreous, resinous, greasy, pearly, silky. Degrees of luster are expressed as splendent, shining, glistening, glimmering, or dull.

Streak. *Streak* is the color of the powder of a substance. In optical-crystallographic studies, it is rarely significant.

Fusibility. *Fusibility* expresses the degree of resistance of a substance to heat, and ordinarily is determined with a blowpipe. Elongate fragments about two millimeters in diameter are held in a blowpipe flame, and the ease of fusion is noted.

Fusibilities are compared with a standard scale, which follows:

1. Stibnite: fused easily in a candle flame.
2. Natrolite (or chalcopyrite): fused easily in a blowpipe flame.
3. Almandite: edges quickly rounded in blowpipe flame.
4. Actinolite: thin edges easily fused in blowpipe flame.
5. Orthoclase: thin edges or splinters fused with difficulty.
6. Bronzite: infusible in blowpipe flame.

Chapter III

ELEMENTARY OPTICS

The Nature of Light. Light is a form of radiant energy. The exact nature of light is not fully understood, but it is known that light constitutes the visible portion of a spectrum extending from high-frequency gamma radiation at one end to low-frequency radio waves at the other end. Just beyond the range of visibility are ultraviolet radiation, of higher frequency, and infrared radiation, of lower frequency.

The process of emission of energy from a substance is called *radiation,* and the emitted energy is also described as *radiation.* Maxwell regarded light as consisting of electromagnetic ether waves of superimposed electric and magnetic fields transmitted with a velocity of approximately 186,000 miles a second. His theory, brought up to date, is called the *electromagnetic theory of light;* it is supported by a vast amount of experimentation and calculation leading to the conclusion that light actually travels through space in continuous waves. However, analysis of radiant energy in terms of the *quantum theory* has yielded an impressive body of evidence that seems to be in direct contradiction to the electromagnetic theory. According to the quantum theory, light travels through space as discontinuous, indivisible particles or bundles of energy called *quanta* or *photons.* In the face of a dilemma, modern theory combines the electromagnetic-wave concept and the particle concept and recognizes them as not necessarily contradictory but as complementary.

The theory of optical crystallography can be developed cogently and consistently if light is regarded as an electromagnetic-wave phenomenon. In this book light is treated as continuous wave motion, and all constructions and derivations are based on this fundamental if somewhat inadequate concept. Radiation of all types is treated as comprising a gradational spectrum which can be analyzed in terms of wave lengths and frequencies.

The following equation is basic:

$$v = f\lambda$$

where v is the velocity of propagation of energy waves in vacuum, a value near 3.0×10^{10} cm/sec (approximately 186,000 miles per sec-

ond); f is the frequency, expressed in cycles per second; and λ is the wave length. Figure 1 shows wave lengths and corresponding frequencies of waves in the electromagnetic spectrum. Note that visible light waves comprise a very limited segment of the spectrum and have

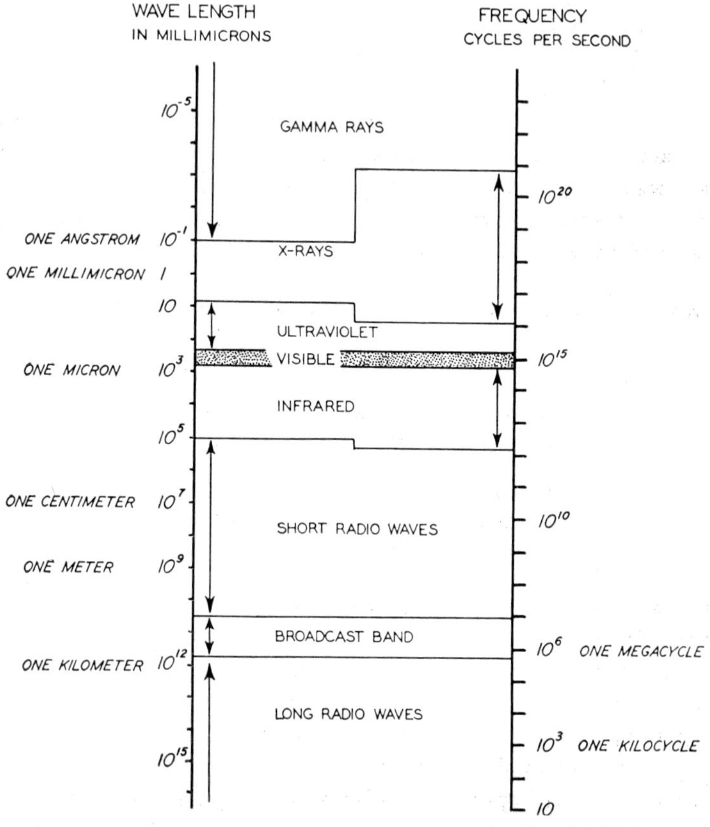

FIG. 1. The electromagnetic spectrum. Diagram shows wave lengths and corresponding frequencies of waves in the interval extending from short gamma rays at one end to long radio waves at the other end of the spectrum.

wave lengths of the order of 400 to 700 millimicrons (approximately 16 to 32 millionths of an inch).

In optical crystallography, wave lengths are commonly expressed in terms of millimicrons, but occasionally one may encounter other units. Following is an expression of equivalence among several units of linear measurement.

$$1 \text{ millimicron } (m\mu) = 10 \text{ angstroms (A)} = 10^{-3} \text{ micron } (\mu) = 10^{-7} \text{ centimeter} = 10^{-9} \text{ meter} = 3.937 \times 10^{-8} \text{ inch.}$$

The Visible Spectrum. White light embraces a gradational series of wave lengths. At one extreme is violet light with a minimum wave length of about 390 millimicrons; at the other extreme is red light with a maximum wave length near 770 millimicrons. Many devices are available for separating white light into its component colors. A beam of white light passed through a triangular glass prism or a diffraction grating is resolved into a color spectrum which is red at one end and grades through orange, yellow, green, blue, and indigo to violet at the other end. Measurement of the wave lengths by means of diffraction gratings or similar equipment gives the range and average wave lengths for the more important components of white light shown in the accompanying table. Light of a particular wave length is described as *monochromatic light.*

WAVE LENGTHS OF LIGHT IN THE VISIBLE SPECTRUM
(in millimicrons)

	Range	*Average*
Violet	390–430	410
Indigo	430–460	445
Blue	460–500	480
Green	500–570	535
Yellow	570–590	580
Orange	590–650	620
Red	650–770	710

Rays and Waves. A *ray* defines the path that light follows in traveling from one point to another. In another sense a *ray* is the light that travels along a line joining two points. In a homogeneous substance or vacuum, light rays are straight lines. However, light moving along rays may be bent or refracted in passing from one substance to another of different properties. A stream of light shining through a very small opening is called a *beam,* and a beam may be considered as a *bundle of rays.*

The expression *light wave* has a double meaning. In one sense, a light wave is the energy emitted from a point source at a given instant and propagated along the path of a ray, advancing by forward motion in combination with a more or less complicated oscillatory motion at right angles to the direction of propagation. The simplest wave of this type is sinusoidal and is the resultant of forward motion combined with simple harmonic oscillation at right angles to the direction of propagation. In a second sense, a light wave is considered as being analogous to a sound wave or a water wave, and the term defines the movement of all of the light energy generated at any instant away from a point source or a source of any size or shape.

For example, in Fig. 2*A*, a point source is emitting light which is traveling outward in all directions with equal velocity in a homogeneous medium. The light emitted at a given instant can be thought of as moving away from the point source as a spherical wave or, in two dimensions, as a circular wave (solid circle). At successive intervals additional waves may be generated as shown by the dashed circles in Fig. 2*A*. In Fig. 2*B* light waves are being generated by a flat surface perpendicular to the plane of the drawing. A series of plane parallel waves, spaced at regular intervals, move away from and parallel to the emitting surface.

Closely related to this concept of light waves is the concept of a *wave front.* If the light waves are pulsating or vibrating systematically or repetitively, points in the waves that are in comparable positions both in space and time can be said to be *in the same phase.* For example, sinusoidal waves are in phase if, after an instant of time, the crests and troughs are in the same relative or actual positions. A wave front may be defined as a surface passing through all points in waves which are in the same phase. Thus, in Figs. 2*A* and *B*, the circles and lines which were drawn to indicate the movement of waves can be used to equal advantage to show a series of wave fronts, each of which passes through all points in the waves in like phase.

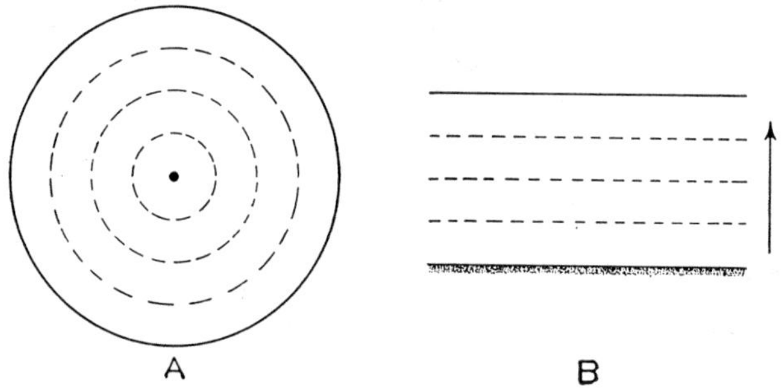

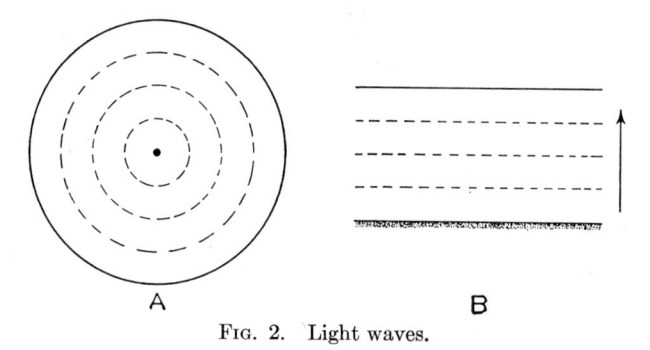

FIG. 2. Light waves.

A. Light waves generated by a point source in a homogeneous medium.
B. Light waves generated by a flat surface and traveling through a homogeneous medium.

The surface to which light has spread along the rays in a given time interval is called the *ray velocity surface* or *ray surface.* In substances in which light travels with equal velocity in all directions, the ray velocity surface derived from a point source is a sphere, the radii of which are the rays. This sphere coincides exactly with the wave front for light from the point source.

In Fig. 3A, O and O' are point sources of light in a medium which transmits light with equal velocity in all directions. Suppose that in a given instant each point has generated in the plane of the drawing a circular ray velocity surface such as the one about O. The radii of the circles are the rays. If consideration is given to the rays in a particular direction, such as OR or $O'R'$, certain relationships between rays and waves become apparent. The light moving along OR and $O'R'$ travels as a wave, and the wave front, which passes through all points in like phase, is obtained by drawing the tangent FF' to the ray velocity surface at R. The wave front FF' indicates the position

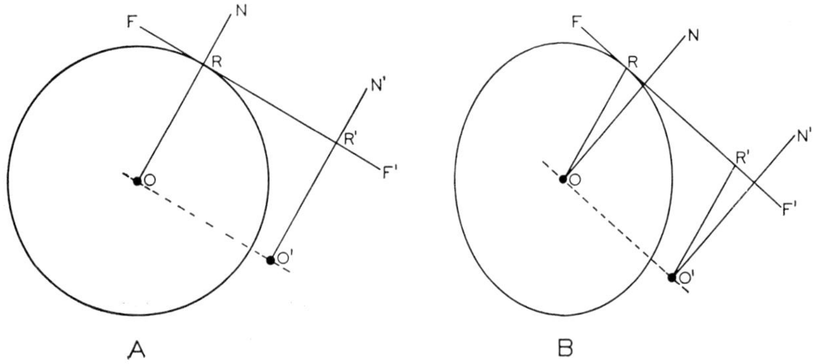

FIG. 3. Ray velocity surfaces.

A. Ray velocity surface in an isotropic medium.
B. Ray velocity surface in an anisotropic medium.

to which a light wave generated at O and O' and moving in the direction OR has reached in the same instant that light moving from O to R along the ray has reached R.

The *wave normal* RN, perpendicular to the wave front, is the direction of propagation of the wave corresponding to the rays OR and $O'R'$ and is parallel to the rays. That is, in media which transmit light with equal velocity in all directions, each ray and its associated wave normal are coincident.

Substances in which light travels with the same velocity in all directions are described as *isotropic.* Vacuum and unstrained gases, liquids, or glassy substances are isotropic. The only crystalline materials that are isotropic are those that belong in the isometric system.

In many materials light travels with different velocities in different directions. This is true of crystals in the tetragonal, hexagonal, orthorhombic, monoclinic, and triclinic systems and in strained isometric crystals or strained noncrystalline materials. Substances in which the velocity of light varies with the direction of transmission are described as *anisotropic.* In Fig. 3B, O and O' are point sources of light in an

anisotropic crystal. In a given interval of time, light traveling along all possible rays reaches surfaces (the ray velocity surfaces) which, in the two dimensions of the drawing, are ellipses such as the one about O. For light traveling along rays OR and $O'R'$, in a particular direction, the *wave front* FF' is obtained by drawing a tangent to the ellipse at point R where the ray OR intersects it. The *wave normal* is ON, perpendicular to the wave front. Note that whereas the light traveling along the rays moves in one direction, the wave, which moves parallel to its normal, travels in a different direction. Thus, for anisotropic crystals it may be said that, in general, waves and the light moving along rays travel in different directions and with different velocities. As for isotropic substances, the ray velocity surface is sometimes designated as the *ray surface.*

Sinusoidal Wave Motion. Wave motion in its simplest form consists of a combination of uniform forward motion and simple-harmonic oscillation at right angles to the direction of forward motion.

Simple-harmonic oscillation is understood by examination of Fig. 4. Imagine a point moving on a circle at a uniform rate in the direction shown by the arrow. Now, at regular time intervals, project the point onto the diameter AM. The projections of A, B, C, etc., so obtained are A, b, c, etc. The oscillatory movement of the projection of the point on the diameter AM is simple harmonic motion. This motion has a velocity of zero at A and M, the points of reversal, and a maximum velocity as it passes through O, the center of the circle. The *amplitude* of the vibration is OA or OM. That is, it is the distance from the center to either point of reversal.

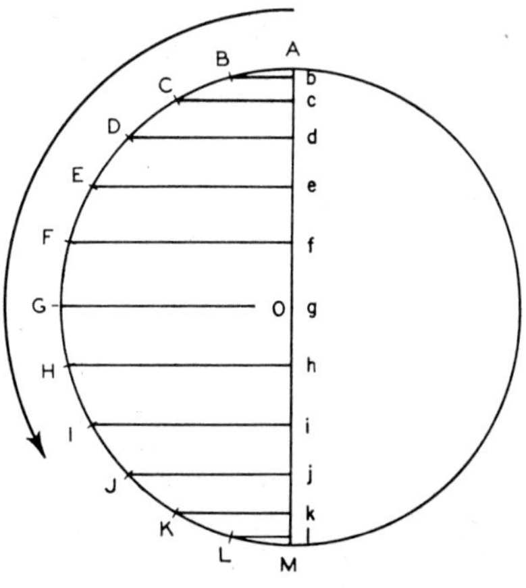

FIG. 4. Diagram illustrating simple harmonic motion.

The *period* is the time consumed in a complete backward and forward motion of the point. For example, in Fig. 4 the period is equal to the time that it takes for the point moving along AM to travel from A to M and back to A.

The nature of wave motion is illustrated in Fig. 5. Imagine a wave, originating at O, moving in the direction OU. The wave consists of the resultant of steady forward motion toward U and simple harmonic vibration parallel to AM. The resultant wave is $ANPQRSTU$.

The *wave length* is NS and may be described as the distance in the di-

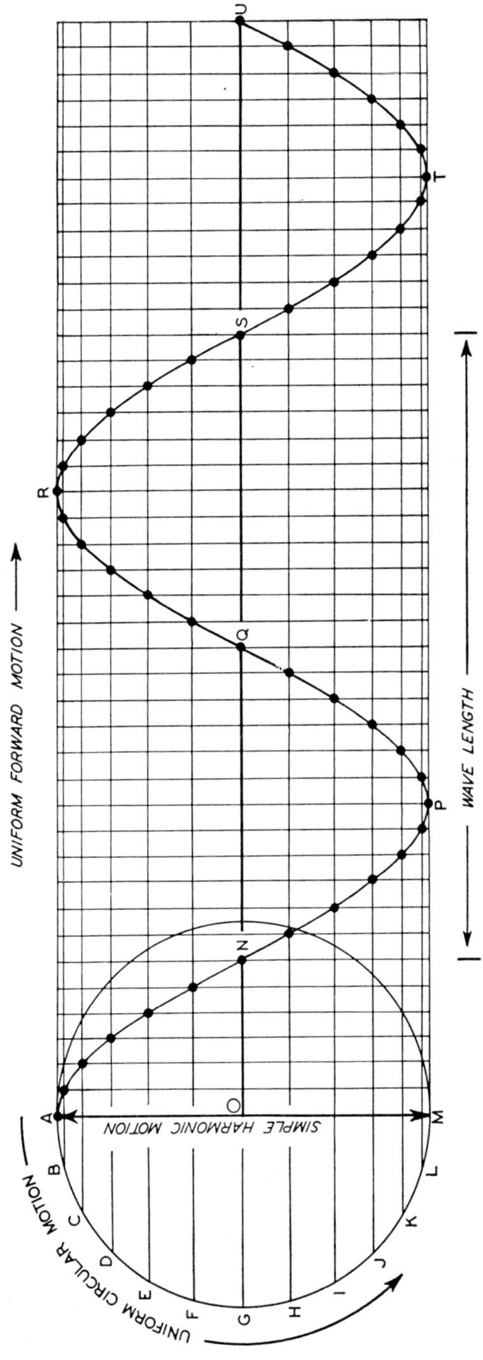

FIG. 5. Simple sinusoidal wave motion.

rection of propagation between points on the wave in like phase. It is designated by the Greek letter λ. *OA* is the amplitude of the wave, and the time that it takes for the wave to travel from *N* to *S* is the *period* of the wave. In light waves the intensity is directly proportional to the square of the amplitude. The number of wave lengths that pass a given point in a given time interval is the *frequency*. Light vibrating in a single plane, as in Fig. 5, is said to be *plane polarized,* and the

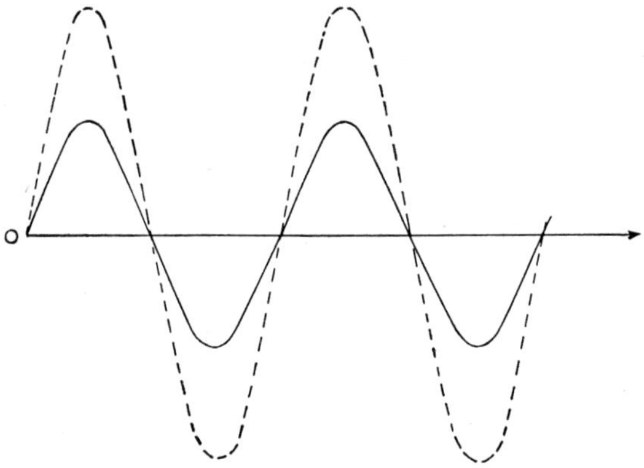

FIG. 6. Constructive interference of two waves of equal phase, wave length, and amplitude. Resultant, dashed line.

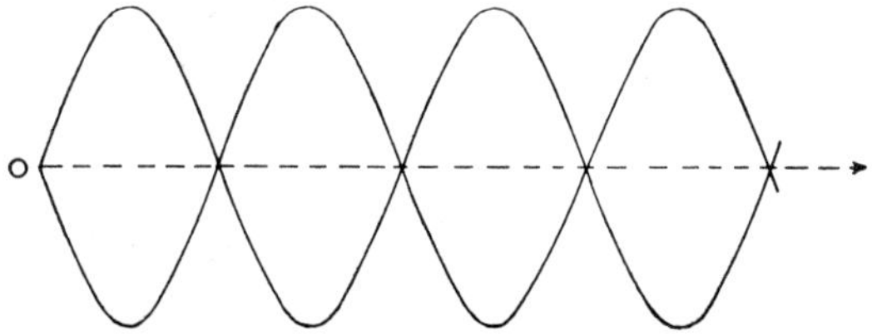

FIG. 7. Destructive interference of two waves with the same wave length and amplitude but with $\frac{1}{2}\lambda$ phasal difference.

plane in which the light vibrates is called the *plane of vibration.* The intersection of the plane of vibration with any surface such as a crystal face gives the *trace* of the plane of vibration in that surface.

Composition and Resolution of Wave Motion. Light waves moving in the same direction along the same path interfere. Figure 6 shows the results of superimposition of one wave on another of like phase, wave length, and amplitude. The resultant wave (dashed line) has twice the amplitude of the original waves and has the same wave length. This illustrates *constructive interference.* Figure 7 shows the interference of two waves of the same wave length and amplitude but

with one-half-wave-length phasal difference. The resultant (dashed line) is zero and shows *destructive interference.* Figure 8 portrays the interference of two like waves having a phasal difference, or path difference, of one-quarter wave length. A general case is illustrated in Fig. 9. Here two unlike waves interfere and produce a wave of irregular amplitude and wave length.

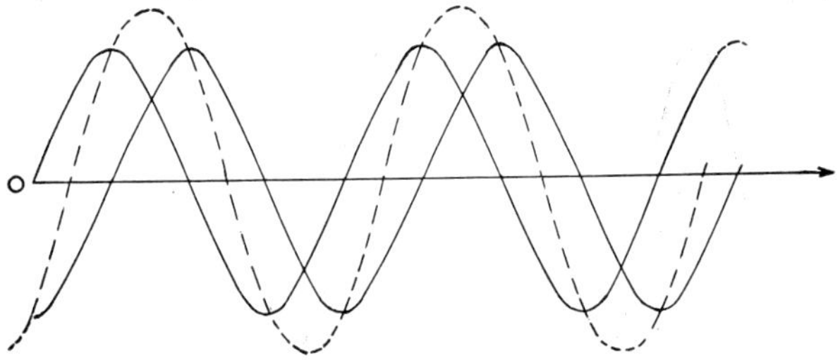

FIG. 8. Interference of two similar waves with $\frac{1}{4}\lambda$ phasal difference.

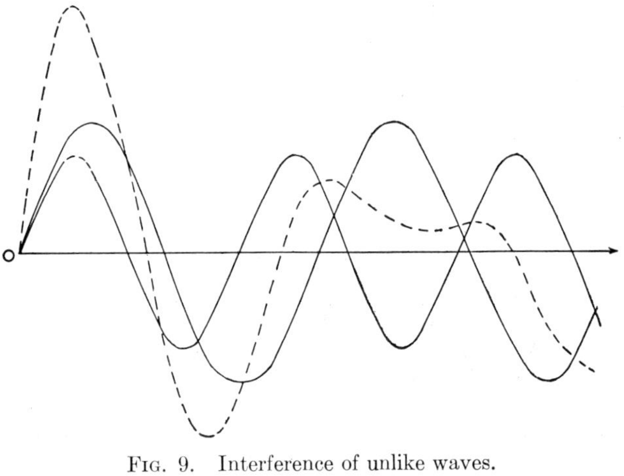

FIG. 9. Interference of unlike waves.

The resultant of the interference of two or more waves vibrating in the same plane and traveling in the same direction is constructed by algebraically adding the perpendicular distances to the waves at all points along the line of transmission.

Figure 10 illustrates the interference of similar waves vibrating in mutually perpendicular planes. The direction of transmission is from right to left. Two identical sinusoidal light waves traveling along the same path interfere in the plane of composition (stippled) yielding a single wave, the characteristics of which depend on the phasal difference (path difference) of the interfering waves. In Fig. 10*A*, the

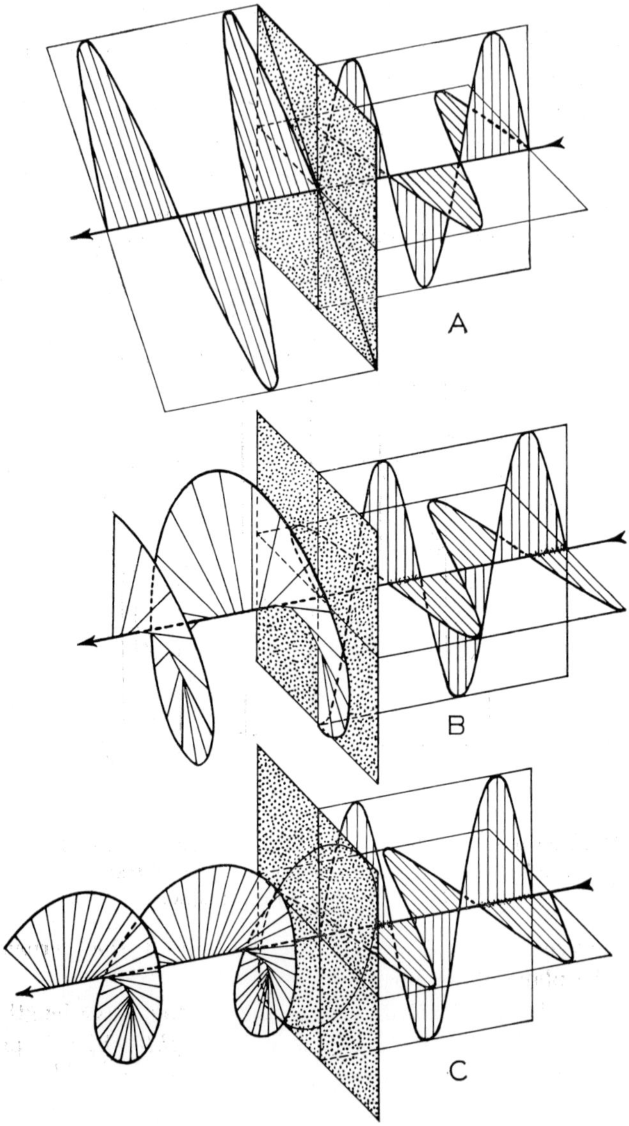

FIG. 10. Composition of waves vibrating in mutually perpendicular planes.

A. Phasal difference (path difference) one-half wave length. Resultant is plane polarized.

B. Phasal difference (path difference) one-quarter wave length. Resultant is circularly polarized.

C. Phasal difference (path difference) one-eighth wave length. Resultant is elliptically polarized.

interfering waves are one-half wave length out of phase, if it is arbitrarily assumed that the displacement of the wave at right angles to its direction of propagation is positive upward for the wave in the vertical plane and positive to the right for the wave in the horizontal plane. The single resultant wave is plane-polarized and vibrates in a plane including a diagonal of the plane of composition. For a phasal difference of one or several whole wave lengths the resultant wave lies in a plane determined by the other diagonal. The amplitude of the

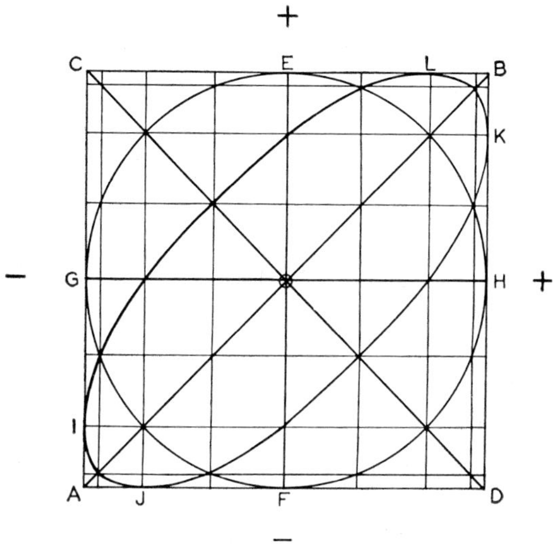

Fig. 11. Diagram used in determining resultant of interference of similar waves vibrating in mutually perpendicular planes. Direction of transmission is perpendicular to the plane of the diagram.

resultant wave is obtained from the vectors of the perpendicular waves in the plane of composition.

Figure 10*B* shows interfering waves one-quarter wave length out of phase. The resultant is *circularly polarized light,* which is produced by mutually perpendicular waves out of phase by $(n/4)\lambda$, where n is 1, 3, 5, 7, etc.

Figure 10*C* illustrates an example in which the interfering waves are out of phase by one-eighth wave length. The resultant wave is *elliptically polarized.* As a matter of fact, plane-polarized light and circularly polarized light may be regarded as special cases, and the statement may be made that the resultant of interference of light waves vibrating in mutually perpendicular planes is, in general, elliptically polarized light.

If the light waves in Fig. 10 travel from left to right, the plane of composition now becomes a plane of resolution, and the complex single wave is resolved into two waves vibrating in mutually perpendicular planes. This phenomenon is observed in light entering anisotropic crystals.

Figure 11 is helpful in explaining Fig. 10. *EF* and *GH* are the traces of the planes of vibration of the mutually perpendicular waves before they interfere. The coordinates are spaced to indicate simple harmonic oscillation of the waves, which are traveling perpendicular to the plane of the drawing. The diagonal *CD*, the circle, and the ellipse correspond to those indicated in the composition plane in Fig. 10.

Finally, in white light, composition of waves of different wave lengths, amplitudes, planes of vibration, and phase produces a motion of a very complex nature.

But no matter how complex the wave motion, it is possible to resolve it into its component wave lengths with a prism or diffraction grating. Moreover, the complex waves of white light are resolved into light waves vibrating in two mutually perpendicular planes in anisotropic crystals.

Chapter IV

OPTICS OF ISOTROPIC SUBSTANCES

Introduction. Isotropic substances transmit light with equal velocity in all directions. The ray velocity surface (ray surface) for a point source in an isotropic medium is a sphere. Geometric constructions showing how light is reflected or refracted are based on *Huygens' principle,* which states that any point or particle excited by the impact of wave energy becomes a new point source of energy. Thus every point in a reflecting surface may be considered as a secondary source of radiation having its own spherical ray velocity surface.

Reflection. A fundamental law of reflection states that the *angles of incidence and reflection, measured from a normal to the reflecting surface, are equal and lie in the same plane.* This plane is called the *plane of incidence.* Figure 1*A* shows light in several parallel rays falling with inclined incidence on a reflecting surface gg'. The wave front for the light in these rays is bb'. The light in each ray sets up a secondary point source of radiation with a spherical ray velocity surface (circles in the plane of the diagram). For example, light traveling along $a'b'$ sets up a new point source at b , and while light in ray abc is traveling from b to c, the light emitted from b' travels out to a ray velocity surface with radius equal to bc. Light moving along other rays behaves similarly, so that it becomes possible to determine the wave front for the light in the reflected rays by drawing a line from c tangent to the ray velocity surfaces generated in the reflecting surface by the points of impact of the light in each incident ray. The wave front obtained in this manner is cc'. The rays for the reflected light are drawn perpendicular to this wave front and are parallel to the wave normal.

In Fig. 1*B* only two rays for the incident light are shown so that the geometric and trigonometric relationships can be seen more easily. Angle i is the angle of incidence and angle r is the angle of reflection. Both are measured from ff', a normal to the reflecting surface gg'. Right triangles $bb'c$ and $b'c'c$ are similar because, by construction, bc equals $b'c'$. Therefore angles $c'cb'$ and $bb'c$ are equal. Now, angles $bb'c$ and i are equal because their sides are mutually perpendicular, and angles $c'cb'$ and r are equal for the same reason. Hence, angles i and r are equal.

The extent to which light is reflected from a surface depends upon the nature of the surface. If the surface is flat and highly polished, none of the light seems to come from the surface, and the reflected object seems to lie beyond the surface. This type of reflection may be called *regular* reflection. If the reflecting surface is rough, it seems to be a source of the light. This results in *diffuse* reflection.

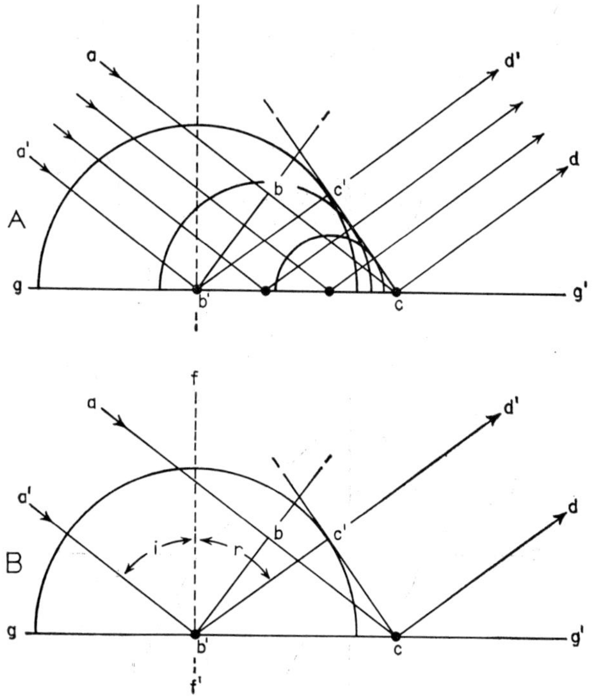

FIG. 1. Reflection by a plane surface.

A. Huygenian construction.
B. Simplified version of *A*.

Luster depends upon the manner and quality of light reflection.

That light penetrates reflecting bodies is shown by the fact that the intensity of reflected light is less than that of the incident light.

Refraction. In general, light passing from one isotropic medium to another is bent or *refracted*. This is not true when the light falls with perpendicular incidence on the contact between the media. In Fig. 2, suppose that gg' is the trace of a plane contact between two substances which transmit light with different velocities. Let V_1 and V_2 be the respective velocities, and suppose that V_1 is greater than V_2; bb' is

the wave front for the light striking the contact gg' with an angle of incidence i. Part of the light is reflected (not shown) and part of the light enters the substance of velocity V_2. The points of impact of light in the rays on the contact gg' serve as point sources of light in medium

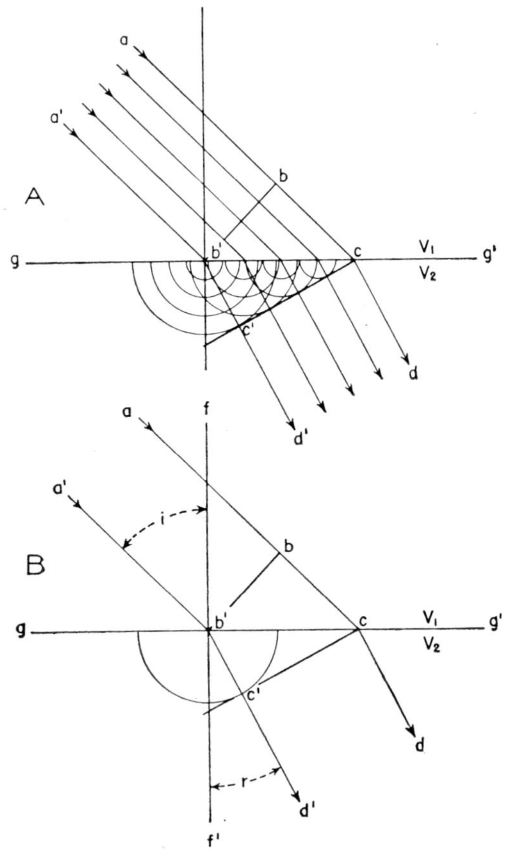

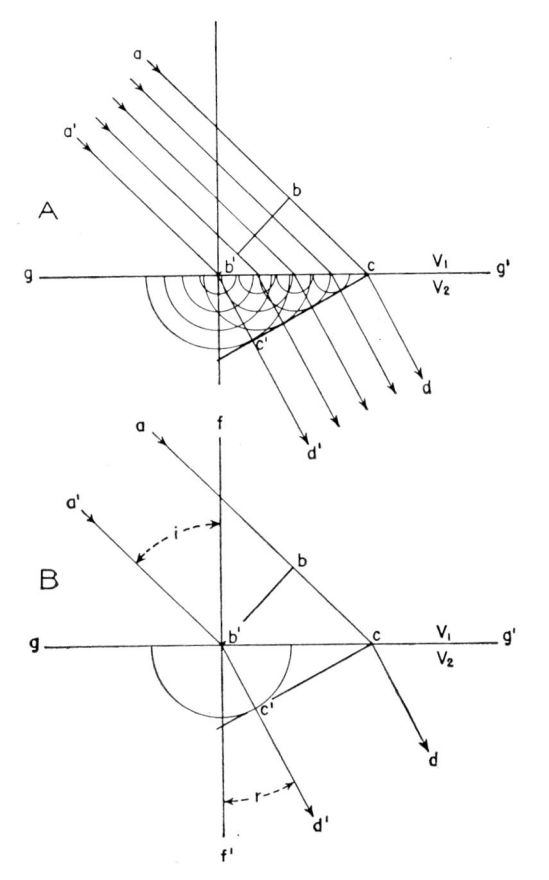

FIG. 2. Refraction of light at a plane surface.

A. Huygenian construction for several rays.
B. Simplified version of *A*.

V_2, so that the wave front in medium V_2 can be obtained by drawing cc' tangent to the ray velocity surfaces for each of the point sources in the plane gg'. The rays in medium V_2 are drawn perpendicular to cc', normal to the wave front.

Figure 2*B* is a simplified version of Fig. 2*A* and shows the important elements of the construction. The ray velocity surface generated by point b' in medium V_2 is drawn with a radius somewhat less

than the distance bc because light travels faster in medium V_1 than in medium V_2. The angle r is the *angle of refraction.*

Index of Refraction. In Fig. 2, by construction $V_1/V_2 = bc/b'c'$, where V_1 and V_2 are the velocities of light in the two media. In right triangles $bb'c$ and $b'c'c$

$$bc = b'c \sin \angle bb'c$$

and

$$b'c' = b'c \sin \angle b'cc'$$

Now

$$\text{angle } bb'c = \text{angle } i \text{ (mutually perpendicular sides)}$$

and

$$\text{angle } b'cc' = \text{angle } r \text{ (mutually perpendicular sides)}$$

so

$$bc = b'c \sin i$$

and

$$b'c' = b'c \sin r$$

Therefore

$$\frac{V_1}{V_2} = \frac{b'c \sin i}{b'c \sin r} = \frac{\sin i}{\sin r} = n, \text{ the } \textit{index of refraction}, \text{ a constant}$$

That is, in isotropic media the ratio between the velocities in the two media is equal to the ratio between the sine of the angle of incidence and the sine of the angle of refraction. This relationship is described as *Snell's law.* The constant n is the *index of refraction* of the medium of light velocity V_2 with respect to the medium of light velocity V_1.

If the angle of incidence is zero, sin i equals zero, and the angle of refraction r becomes zero. That is, light normally incident on a plane surface is not refracted. Light passing obliquely from air into a solid is invariably refracted toward the normal, and light passing from a solid into air is refracted away from the normal. This rule also holds when light passes from an optically rarer to an optically denser medium, or when the reverse is true.

The index of refraction of a substance differs for the various wave lengths of light. This fact can be demonstrated by the familiar prism experiment shown in Fig. 3. A narrow beam of white light falling on the wall of a prism cut from glass or some other transparent substance is split into the individual colors that comprise the visible spectrum; this proves that the prism has a different refractive index for each of the wave lengths. Light of short wave length has a lower velocity and is refracted more than light of longer wave length, thus permitting

the generalization that *the index of refraction varies inversely as the wave length of light.*

The power of a substance to refract light is sometimes described as *refringence.* Substances of high refractive index have high refringence; those of low index, low refringence.

A generalized formula for the index of refraction may be written as

$$n = \frac{c}{v}$$

in which n is the index of refraction, c is a constant, and v is the velocity of light of a particular wave length in a specified medium. The con-

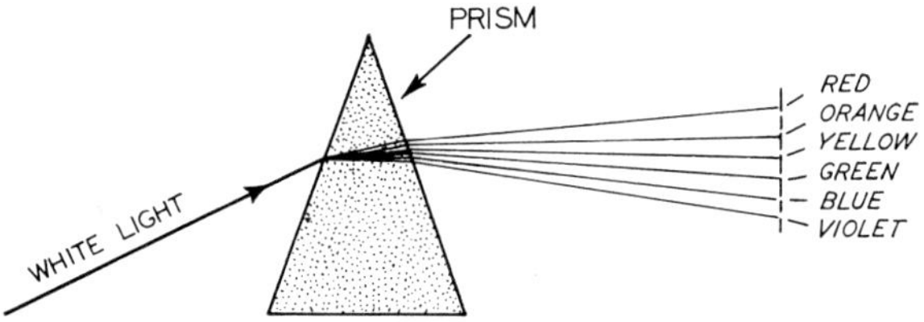

FIG. 3. Formation of color spectrum by a beam of white light passing through a prism.

stant c is generally regarded as the velocity of light in a vacuum and is arbitrarily assigned a value of unity, so that the above equation reduces to

$$n = \frac{1}{v}$$

The average index of refraction of air at sea-level pressure is 1.00029, indicating that the velocity of light in air is only slightly less than its velocity in vacuum. Thus a very small and, for most purposes, a negligible error is introduced in refractive-index measurements in air.

The index of refraction, as defined above, applies without qualification to isotropic substances, but in anisotropic substances having two or three principal refractive indices the definition should be phrased more precisely. In isotropic substances light transmitted along rays moves with the same velocity as light traveling in the direction of the wave normal. However, in anisotropic substances the wave-normal velocity is not necessarily the same as the ray velocity.

In this text, the velocity of light in a substance is regarded as the velocity in the direction of the wave normal unless otherwise specified. In the equation

$$n = \frac{\sin i}{\sin r}$$

i is the angle of incidence in vacuum (or air), and r is the angle of refraction of light traveling along the *wave normal* in the refracting medium.

The Isotropic Indicatrix. If it is assumed that a light wave vibrates at right angles to the direction of its propagation, the passage of light through an isotropic medium can be related to a sphere called an *isotropic indicatrix* (Fig. 4). All radii of the sphere are equal to n, the index of refraction of the medium for a particular wave length; the radii give the refractive indices in the directions of their vibration

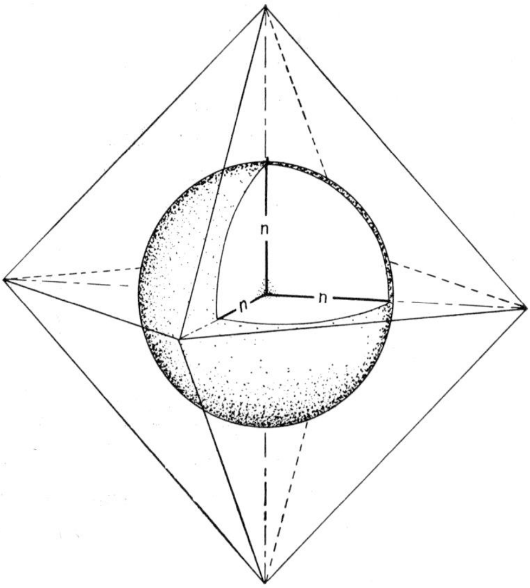

FIG. 4. The isotropic indicatrix.

of all waves originating at a point source. The wave corresponding to a particular radius of the sphere travels at right angles to the radius with a velocity proportional to $1/n$.

Total Reflection and Critical Angle. Light traveling from one medium to another of higher refractive index is refracted toward the normal in the medium of higher index. Conversely, light traveling in the opposite direction is bent away from the normal in the medium of lower index.

Figure 5 shows light passing from a medium of velocity V_2 to a medium of velocity V_1. V_2 is less than V_1. Inasmuch as in the derivation of the formula for the index of refraction i is used to designate the angle of incidence in the medium of higher velocity, it will be retained here with the understanding that the light is actually moving

in the opposite direction, that is, from medium of velocity V_2 into medium of velocity V_1. In Fig. 5*A*, light entering a medium of velocity V_1 is refracted away from the normal *ff′*, and, if V_1 is the velocity in air or vacuum,

$$n = \frac{\sin i}{\sin r}$$

In Fig. 5*B*, the angle r is such that the angle i is 90 degrees; that is, the light entering air just grazes the surface of contact between the two media. The angle r, when this condition obtains, is called the *critical angle.* The following relation now holds:

$$n = \frac{1}{\sin r} \quad (\sin 90° = 1)$$

For example, diamond has a refractive index of 2.42, and 2.42 = 1/sin r, from which it follows that the critical angle is 24° 24′. When the critical angle is exceeded, all of the light is internally reflected as in Fig. 6*C*. This type of reflection is described as *total internal reflection.*

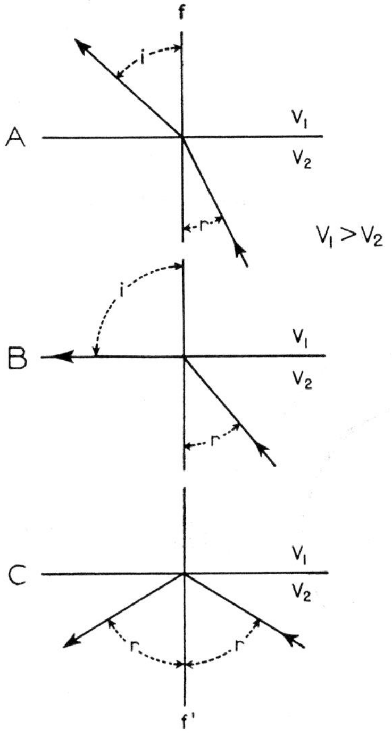

FIG. 5. Refraction and reflection of light at plane contact of two isotropic media.

The refractive indices of two substances are related by the equation

$$\frac{n_1}{n_2} = \frac{\sin i}{\sin r}$$

where n_1 is greater than n_2, and i is measured in the medium of lower index. Thus, a critical angle r can be determined for any substance in contact with another of lower refractive index. At this critical angle

$$\frac{n_1}{n_2} = \frac{1}{\sin r}$$

where r is measured in the medium of higher index. In air or vacuum, n_2 becomes unity.

Lenses. Lenses are optical devices for refracting light and are made of transparent isotropic substances. The surfaces of lenses are generally spherical, and the curvatures of the two surfaces are equal or different depending upon the intended use of the lens. The line passing through the centers of curvature is the *axis* of a lens.

Fig. 6. Simple lenses.

A. Biconvex. *D*. Plano-convex.
B. Biconcave. *E*. Concave-convex.
C. Plano-concave. *F*. Concave-convex.

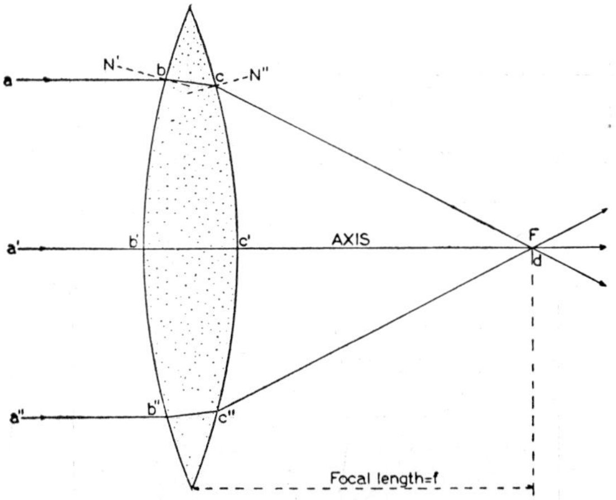

Fig. 7. Passage of light through biconvex lens.

Figure 6 illustrates various types of simple lenses. Depending on the manner in which it refracts light, a lens may be said to be *converging* (positive) or *diverging* (negative). Refraction by a biconvex converging lens is illustrated in Fig. 7. Light moving along a ray *abcd*, incident at *b*, is refracted toward the normal N', and as it leaves the lens is refracted again, this time away from the normal. Light in a ray, $a'b'c'd'$, parallel to the axis, is not refracted. Light in ray $a''b''c''d''$ is refracted so as to intersect refracted light in other rays at *F*, the *focal point* or *focus*. The distance *f* from the center of the lens to the

focal point is termed the *focal distance* or *focal length*. The focus at F is a *real focus*, and the image at that point is a *real image*.

A biconcave lens is shown in Fig. 8. This lens causes light to diverge. The rays for light passing through the lens, if projected backward, intersect at point F, the *virtual focus*, and it is said that a *virtual image* is formed at F.

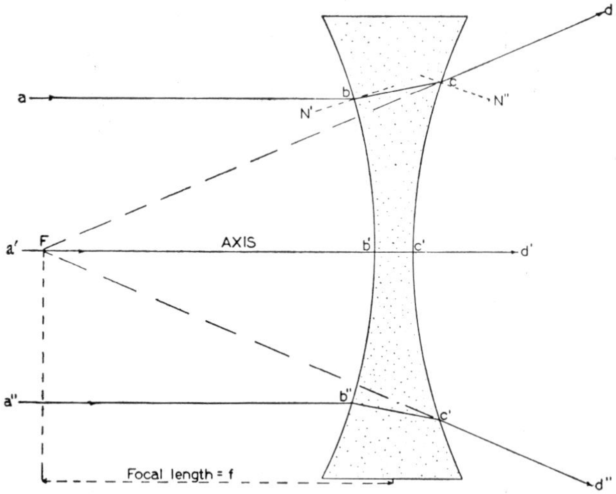

FIG. 8. Passage of light through biconcave lens.

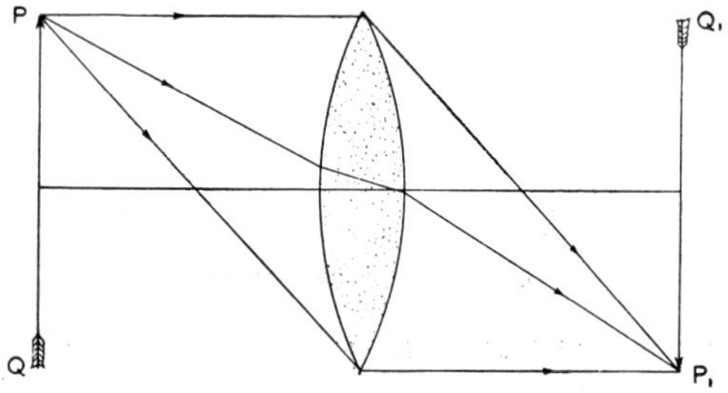

FIG. 9. Reversal of image by a lens.

Lenses commonly reverse images as shown in Fig. 9. Light from the tip of the arrow passes through the lens and comes to a focus on the opposite side in such a way that the image of the tip is reversed. Light in other rays behaves similarly, resulting in the reversal of the image of the whole arrow.

Defects of Lenses. Lenses suffer from certain inherent defects called *aberrations*. *Chromatic aberrations* result from the fact that a lens

has different refractive indices for the component colors of white light. Other defects, present even in monochromatic light, are called *monochromatic aberrations.*

Spherical aberration is illustrated in Fig. 10. The point P yields an image at P' or P'', depending upon the path followed by the light through the lens. Light passing through the central portion of the lens produces an image at P', and that passing through the lens near its edge forms an image at P''. Evidently, there is no single plane in which a sharp image of P will be formed. This type of aberration

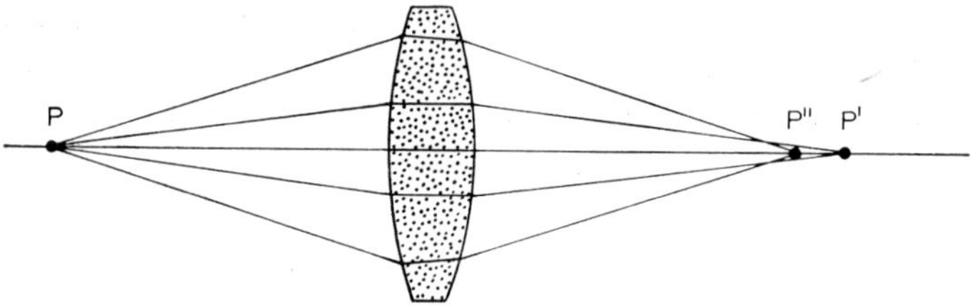

FIG. 10. Spherical aberration in a simple lens.

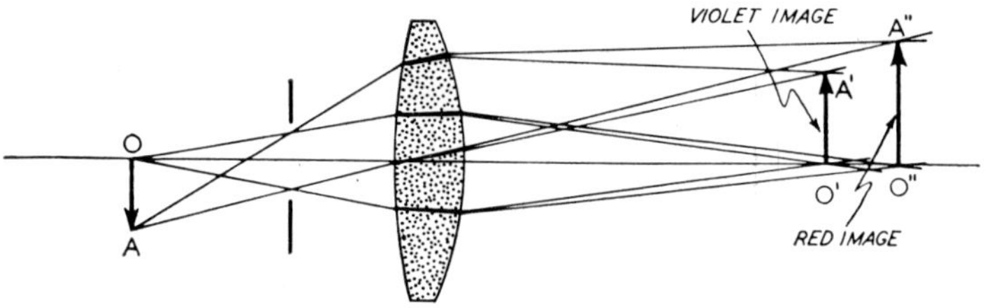

FIG. 11. Chromatic aberration in a simple lens.

is reduced by employing a diaphragm so that only the central portion of the lens transmits light.

Chromatic aberration, shown in Fig. 11, results from the fact that the lens has different refractive indices for different wave lengths of light. In white light, an arrow OA will produce a series of reversed images corresponding to each of the colors of the visible spectrum. The images $O'A'$ for violet and $O''A''$ for red are at the extremes of the spectrum and indicate the maximum limits of aberration. Note in Fig. 11 that not only are the images of the arrow corresponding to the various colors at different focal distances, but they are of different size.

Other types of aberrations are *coma, astigmatism,* and *distortion. Coma* affects the image of a point not lying in the axis of a lens and tends to spread the image out over a plane at right angles to the

lens axis. *Astigmatism* affects the image of a point not on the lens axis and tends to spread the image in a direction along the axis. *Distortion* arises not from a lack of sharpness of the image but from a variation of magnification with axial distance.

Balancing of Aberrations. The inherent defects of simple lenses may be partly or completely nullified by constructing *compound lenses.* A compound lens consists of a number of individual lenses assembled in such a manner that the aberrations of one part of the system are balanced against those of another part. However, as a practical matter, it is almost impossible to construct a compound lens which overcomes all of the various types of aberrations; so, in general, a lens is constructed in such a manner as to remove or considerably reduce the aberrations that are detrimental to a specific use.

Chapter V

THE POLARIZING MICROSCOPE

Introduction. The polarizing microscope originally was constructed for the petrographic examination of thin slices of rocks, but in recent years it has assumed increasing importance in the fields of chemistry, metallography, and medicine. Many branches of industry make advantageous use of the polarizing microscope. Because of widely varying applications, the polarizing microscope has undergone many modifications, but in principle all types do not differ essentially from the original petrographic microscope. Accordingly, the descriptions here will be limited to the type of microscope ordinarily employed by the petrographer or the chemical microscopist.

The polarizing microscope will perform all of the functions of an ordinary microscope. Moreover, it permits the evaluation of properties and characteristics that cannot be measured by other means. A typical polarizing microscope differs from an ordinary microscope in that it possesses a revolving graduated stage and two polarizing prisms, one below the stage and one above. In addition, a removable, auxiliary Bertrand-Amici lens is present between the upper polarizing prism and the eyepiece. Figures 1 and 2 illustrate two types of polarizing microscopes.

Construction of the Polarizing Microscope. The microscope may be made to serve a twofold purpose. If the Bertrand-Amici lens is not inserted, the microscope acts as an *orthoscope,* and the optical system magnifies an object on the stage. The paths of the light in an orthoscope are indicated in Fig. 3.

When the Bertrand-Amici lens is inserted, the microscope is converted into a *conoscope* and is used for the examination of magnified interference phenomena. The Bertrand-Amici lens in effect brings the eyepiece into focus on a focal plane of the objective and brings the entire optical system to a focus at infinity. Ordinarily, for the examination of interference figures, it is necessary to use a high-power objective and a strongly converging substage condensing lens. The paths of light, when the microscope is used as a conoscope, are indicated in Fig. 4.

Objectives. Microscope objectives produce a magnified image of an object on the microscope stage. The ratio of the size of this image

to the size of the object is termed the *initial magnification of the objective.* The magnification increases with an increase in the tube

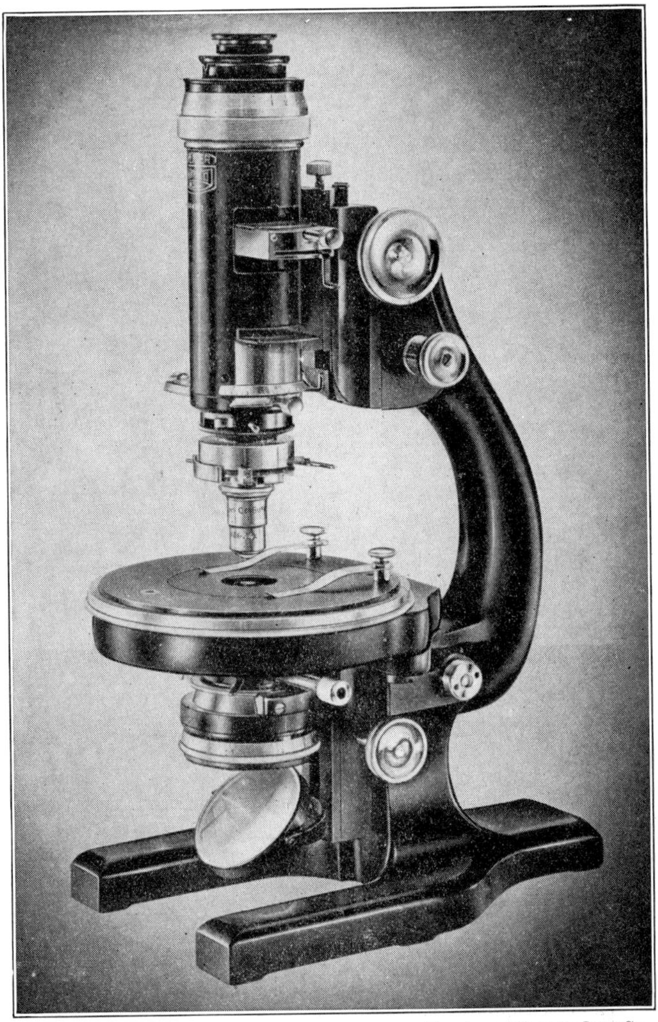

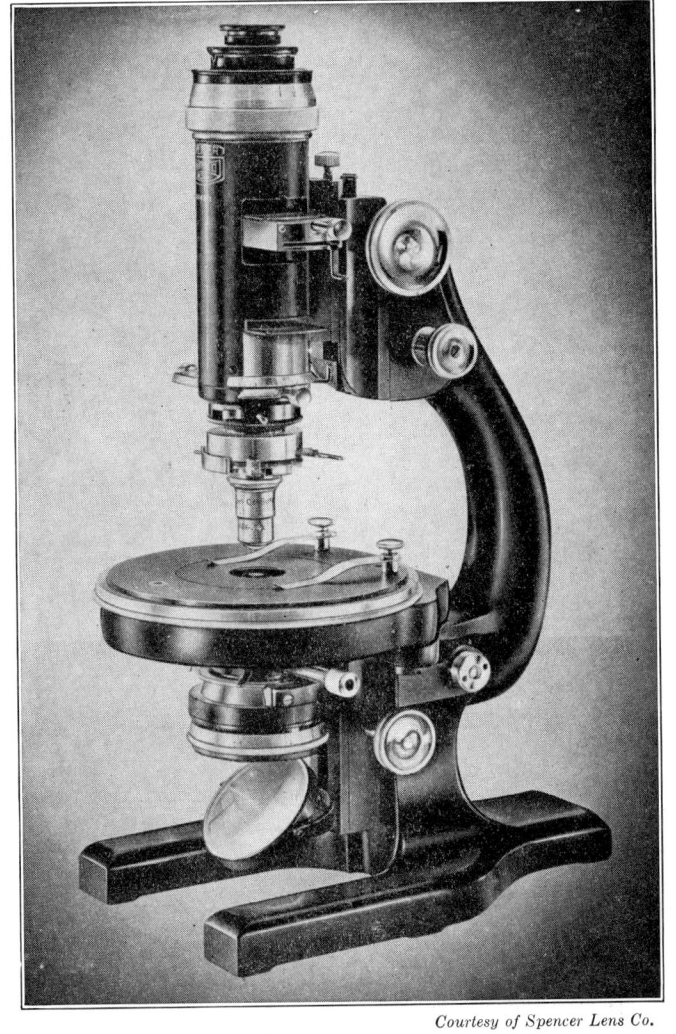

Courtesy of Spencer Lens Co.

FIG. 1. Petrographic microscope.

length and with a decrease in the focal length of the objective lens system.

The *resolving power* of an objective is an important property which measures its ability to show as distinctly separated in the structure of an object two small elements which are a small distance apart.

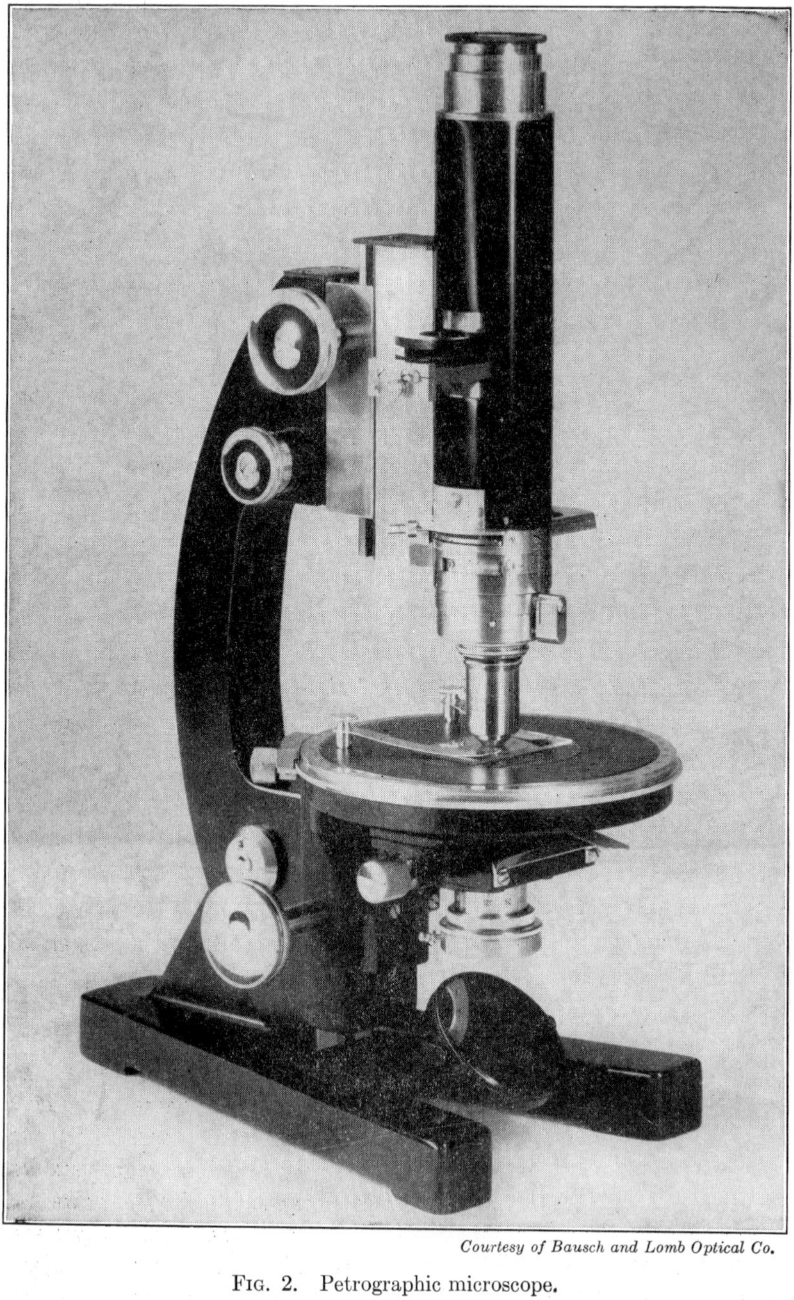

Courtesy of Bausch and Lomb Optical Co.

FIG. 2. Petrographic microscope.

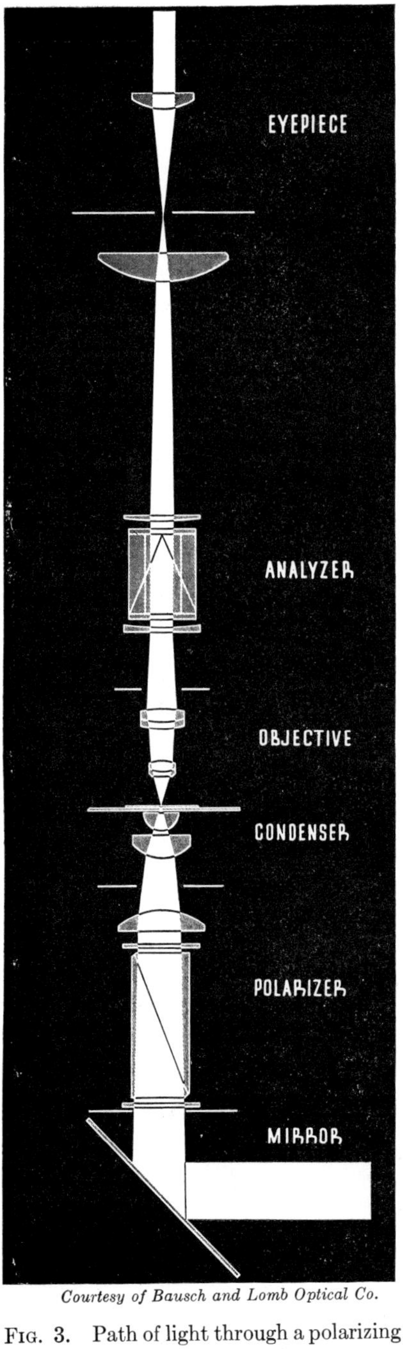

Courtesy of Bausch and Lomb Optical Co.

FIG. 3. Path of light through a polarizing microscope.

Resolving power is a function of the numerical aperture (N.A.) and increases with an increase in the numerical aperture. The following relationship holds:

$$N.A. = n \sin u$$

where n is the lowest refractive index between the object and the objective, and u is half the angular aperture of the objective. With most objectives n is the index of air, which is slightly greater than unity. The angular aperture measures the angle between the directions followed by light passing through the objective with maximum obliquity.

The application of the principle discussed above is well illustrated by the oil-immersion objective. At high magnifications, the resolving power of a lens may be increased by displacing the air space between a magnified object and the objective with a liquid layer which has an index of refraction considerably higher than that of air.

The *depth of focus* of an objective depends on the numerical aperture and magnification: it is inversely proportional to both. Thus, as the numerical aperture and magnification increase, the depth of focus decreases. Any effort to increase depth of focus by means of diaphragms results in decreased resolving power.

Various types of objectives are available. Apochromatic objectives are constructed so as to reduce color aberration to a

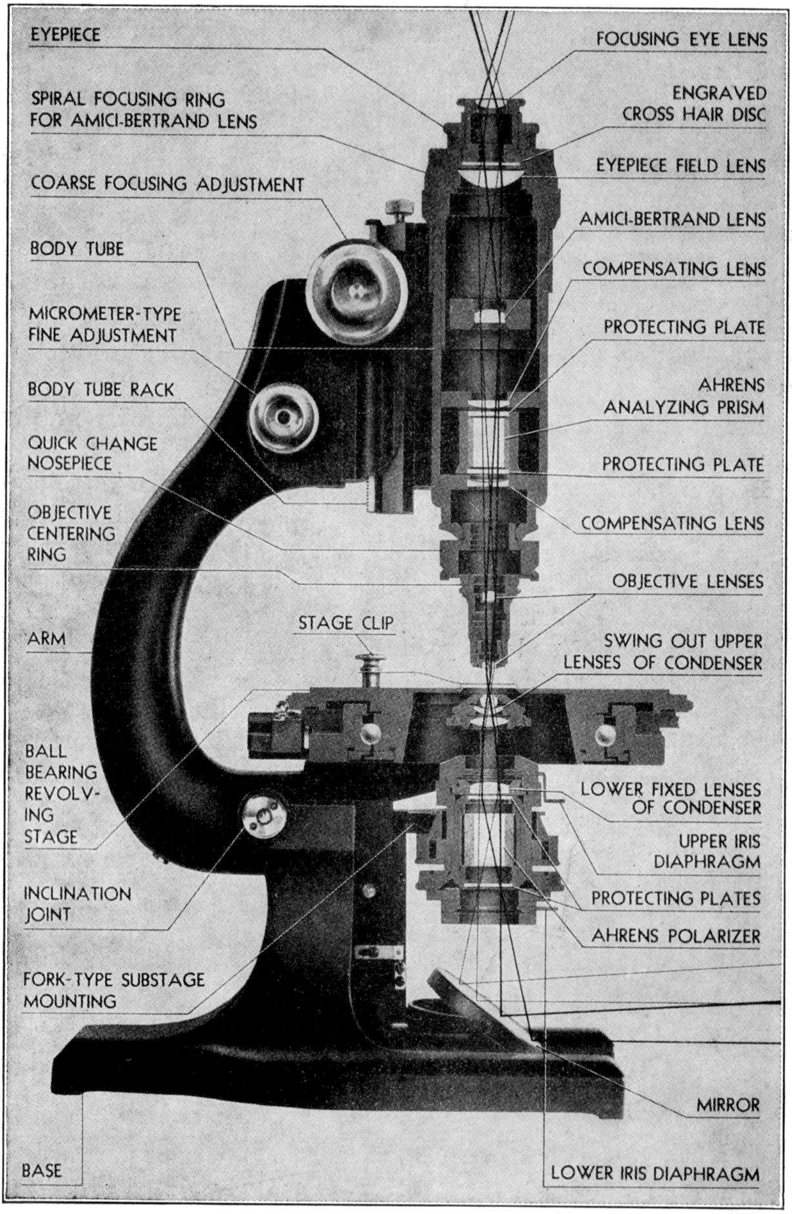

Courtesy of Spencer Lens Co.

FIG. 4. Cross section of petrographic microscope.

minimum. Such lenses, however, may suffer from a decided lack of flatness of field. This condition is corrected in part by careful construction.

Achromatic objectives are not corrected for spherical and chromatic aberration to the same extent as apochromatic objectives but are satisfactory for most purposes. These objectives should be used for exact optical measurements because they do not rotate the plane of polarization of transmitted light as much as other types of objectives.

For special types of work fluorite objectives may be useful. These consist of a combination of glass and fluorite lenses and have better resolving power than achromatic objectives. Fluorite and apochromatic objectives find wide use in photomicrography at high magnifications.

The focal length, numerical aperture, and magnification usually are engraved on the barrel of an objective.

Eyepieces. *Eyepieces,* or *oculars,* serve to magnify the image produced by the objective. Eyepieces are constructed in various ways and with various magnifications. To compute the magnifying power of a microscope all that is necessary is to multiply the magnification of the objective by that of the eyepiece. Thus, if an objective has a magnification of 43 and an eyepiece a magnification of 10, the total magnification is 430×.

Compensating eyepieces are constructed primarily for use with apochromatic objectives. These eyepieces eliminate color fringes which are conspicuous when ordinary eyepieces are used.

The simplest type is the *Huygenian eyepiece,* which is constructed for use with achromatic objectives. *Hyperplane* eyepieces have a color compensation midway between those of the Huygenian and compensating eyepieces; they are useful for high magnification and in photomicrography.

Polarizing Prisms. The polarizing prisms of a petrographic microscope consist of clear calcite, cut and recemented with balsam so as to plane-polarize transmitted light. The lower prism, below the microscope stage, is called the *polarizer;* the upper prism, above the objective, is termed the *analyzer.* For most purposes the planes of polarization of the prisms are kept in mutually perpendicular positions. The theory of polarizing prisms is discussed in Chapter VIII.

In certain types of microscopes now on the market, polarization of light is accomplished by selective absorption in a disc containing a very strongly pleochroic material. The disc permits light vibrating in only one direction to pass through.

Chapter VI

MEASUREMENT OF INDEX OF REFRACTION

Introduction. Measurement of the refractive indices of solids is a very useful procedure in determinative mineralogy or chemistry. There are many methods of index measurement, each suited to a certain purpose and yielding results within certain limits of accuracy. The method most widely employed is the immersion method, in which the index of a solid is compared with that of a liquid of known index. Many variations of the immersion method have been devised: some are adaptable to routine laboratory work with a minimum of equipment; others require expensive and complicated apparatus and are not widely used.

The discussion of index determination presented in this chapter is based on the assumption that the measured substance has only one index of refraction for a particular wave length of light; that is, it is isotropic. It should be understood that uniaxial crystals have two principal indices of refraction and that biaxial crystals have three. The measurement of all indices of refraction of anisotropic crystals is discussed in subsequent chapters.

Relief. Under the microscope, crystal sections or fragments generally are characterized by rough, irregular, sometimes pitted surfaces, which result in the phenomenon of *relief* (or shagreen). The relief of a crystal fragment depends upon the relative indices of the fragment and the medium with which it makes contact. As the indices of the fragment and the surrounding medium approach each other, the relief diminishes.

Relief is an expression of the fact that as light passes from a medium of one index to a medium of a different index, the surface of contact acts so as to refract or totally reflect the light. If a crystal is immersed in a medium of considerably higher or lower index, every imperfection or flaw becomes conspicuous, and the relief is said to be high. Relief is absent when the index of the crystal and the immersion medium are the same, because light is not reflected or refracted at the contact but passes without path deviation through solid and immersion medium.

Relief depends on the *difference* in the indices of a substance and the immersion medium and *not* on the fact that the immersion medium

has an index higher or lower than the substance. However, *apparent relief* may result from inclusions, alteration products, cleavage, internal fractures, or absorption of transmitted light.

The nature of relief is illustrated by the diagrams in Fig. 1. In each example it is assumed that a corrugated glass plate of index N is placed in an immersion medium of known index. In Fig. 1A the index of the immersion medium is considerably less than that of the glass, and, as the light passes from the glass into the immersion medium, it is strongly bent or refracted. In consequence, as indicated

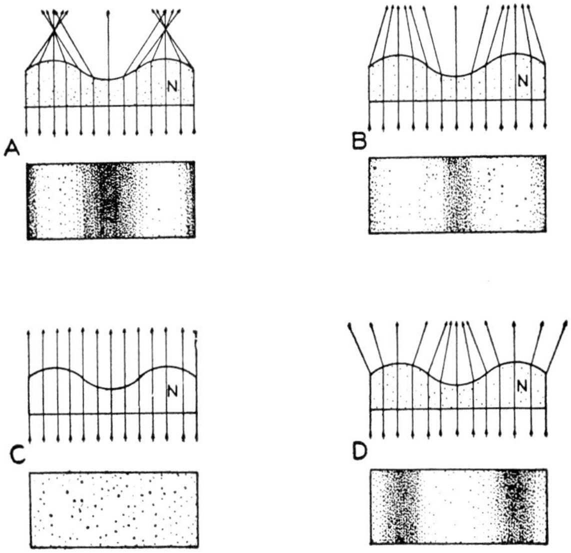

Fig. 1. Relief.

by the top view of the glass plate, the relief is high, and each trough and ridge stands out prominently. In Fig. 1B the index of the immersion medium is slightly less than that of the glass plate, and the relief is lower than in Fig. 1A. In Fig. 1C the indices of the glass plate and the immersion medium are the same, and the relief disappears. In Fig. 1D the index of the immersion medium is considerably higher than that of the glass, and the relief again is high.

The diagrams in Fig. 1 do not show the effect of total reflection of light at the surface of contact between the glass and the immersion medium. However, total reflection contributes to relief in fragments which have curved or rough surfaces.

Becke Line. The *Becke line,* as originally defined, refers to a phenomenon associated with a vertical contact of two substances of differ-

ent indices of refraction observed on the stage of a microscope. The Becke line is seen to best advantage under the microscope when the high-power objective is used or when the medium-power objective is used with reduced illumination.

The light leaving the low-power condensing lens just below the microscope stage is slightly convergent. In Fig. 2 substances of indices n and N make a vertical contact. N is larger than n. Part of the light in rays 1 and 2, passing through the medium of index n and striking the contact between the two media, is reflected, and part is refracted into the medium of index N. The reflected light in medium of index n is not shown in the diagram. Light in rays 3 and 4 passing through the medium of index N is totally reflected. The effect is a concentration of light above the contact on the side toward the medium of higher index.

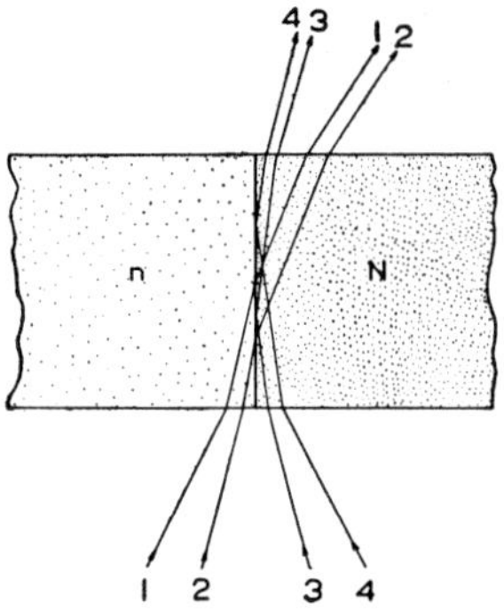

FIG. 2. Origin of the Becke line.

Under the microscope the Becke line is not seen when the microscope is exactly focused on a fragment. However, if the tube of the microscope is slightly raised, a narrow line of light (the Becke line) appears just inside or outside the contact of the fragment with the surrounding medium. If the Becke line moves into the fragment when the microscope tube is raised, the index of the fragment is higher than that of the surrounding substance. If the Becke line moves out into the surrounding medium when the tube is raised, the fragment has a lower index than that of the surrounding medium. Lowering the tube reverses the effect in each instance.

The lateral movement of the Becke line as the microscope tube is raised or lowered beyond the position of clear focus on an object on the microscope stage is explained by reference to the diagrams in Fig. 3. In Fig. 3*A* the objective lens is focused on the vertical contact between two substances of different refractive indices, n and N; n is smaller than N. The light totally reflected on the side of the contact toward substance of higher index N comes to a focus in the focal plane above the lens, and the observer is not aware of a concentration of light. He sees only the magnified image of the object on the microscope stage. In Fig. 3*B* the microscope tube is raised above the position of sharp focus, and the focal plane of the lens is elevated above the image formed by the contact. In this position the observer can still observe the contact between the two substances, but the

image is slightly out of focus. Moreover, the light concentrated in substance of index N forms an image below the upper focal plane of the lens, and the Becke line makes its appearance on the side of the contact toward the medium of higher index. In Fig. 3C lowering the objective lens below a sharp focus produces an opposite effect: the Becke line moves into the medium of lower refractive index.

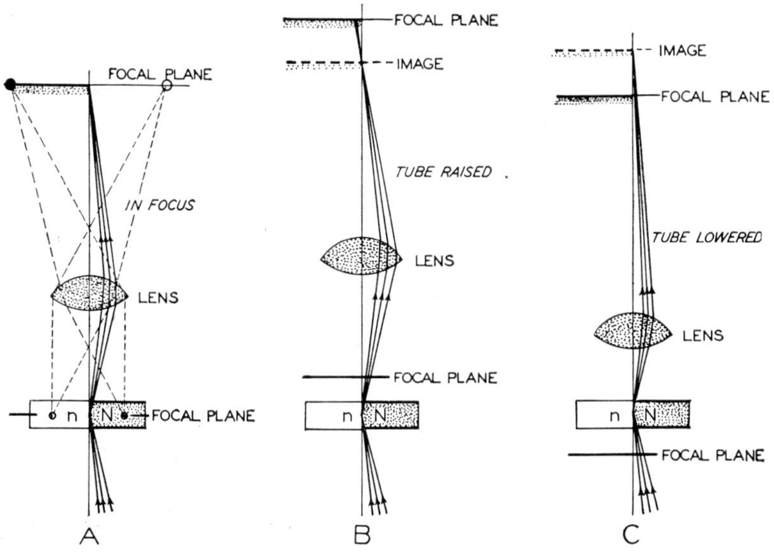

Fig. 3. Diagram illustrating movement of Becke line as microscope tube is raised or lowered from the position of sharp focus.

A. Objective lens focused on object on microscope stage.
B. Objective lens raised above position of sharp focus.
C. Objective lens lowered below position of sharp focus.

Central Illumination. Fragments of solids commonly are crudely lenticular in cross section, and, if embedded in an immersion medium, act as biconvex lenses. The effect of refraction of light under such circumstances is diagrammatically indicated in Fig. 4. In Fig. 4A it is supposed that the fragment is immersed in a medium of higher index. Such a condition results in the dispersal of light passing through the fragment. In Fig. 4B the fragment has a higher index than that of the medium in which it is immersed. The refraction of light produces a concentration of light above the fragment.

By considering central illumination alone, the index of a fragment relative to the medium in which it is immersed may be determined by raising the microscope tube slightly above the position of clear focus.

If its index is higher than that of the immersion medium, the central portion of the fragment is illuminated when the tube is raised. If the fragment has a lower index, its central portion is darkened.

Ordinarily, central illumination is observed when the field of the microscope is somewhat darkened by the use of a substage diaphragm.

In practice crystal fragments in immersion media show a composite effect. Both central illumination and the Becke line play a part, and each supplements and reinforces the other.

Central illumination as related to the position of the objective lens and a fragment on the microscope stage is shown in Fig. 5. A lenticular fragment of higher refractive index than the surrounding medium causes a concentration of light above the fragment. In Fig. 5*A* the focal plane of the lens coincides with the fragment, and the image is in sharp focus. The observer is not aware of the concentration of light above the fragment until the tube of the microscope is raised as in Fig. 5*B*. Lowering the objective lens below a position of sharp focus, as in Fig. 5*C*, causes the central portion of the fragment to appear to darken.

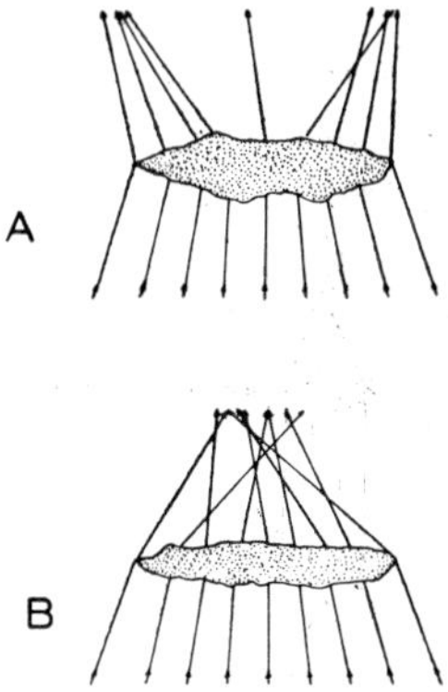

FIG. 4. Refraction of light by lenticular fragments.
A. Fragment has lower index than surrounding medium.
B. Fragment has higher index than surrounding medium.

Figure 6 shows diagrammatically the composite effect of central illumination and the Becke line. In Fig. 6*A* the microscope is exactly focused on a fragment immersed in oil. The field is darkened by the use of a substage diaphragm. Figure 6*B* illustrates the result of raising the microscope tube when the grain has a higher index than that of the immersion medium. The center of the grain is illuminated and a fine line of light (the Becke line) moves from the edge into the grain. In Fig. 6*C* it is assumed that the fragment has a lower index than the material in which it is embedded. When the microscope tube is raised, the center of the fragment is darkened, and the Becke line moves into the immersion medium. The arrows indicate the direction of movement of the Becke line in the two examples.

If the microscope tube is lowered below a clear focus, the effects are the opposite of those just described.

Oblique Illumination. The relative indices of a substance and an immersion medium may be determined satisfactorily by the *method*

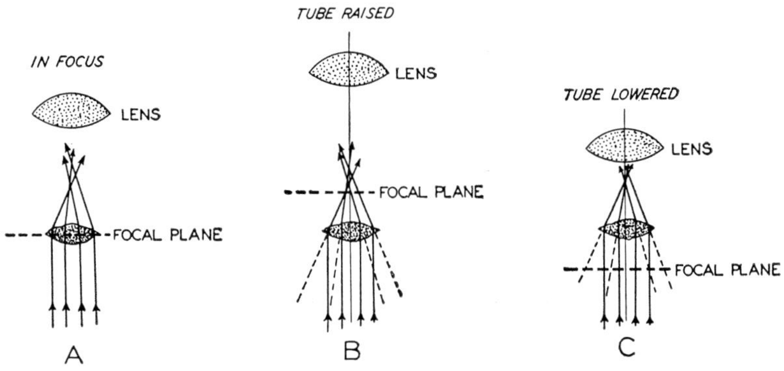

Fig. 5. Central illumination as related to the position of the objective lens of a microscope. Fragment has a higher index of refraction than surrounding medium.

A. Lens in sharp focus on object on microscope stage.
B. Objective lens raised above position of sharp focus.
C. Objective lens lowered below position of sharp focus.

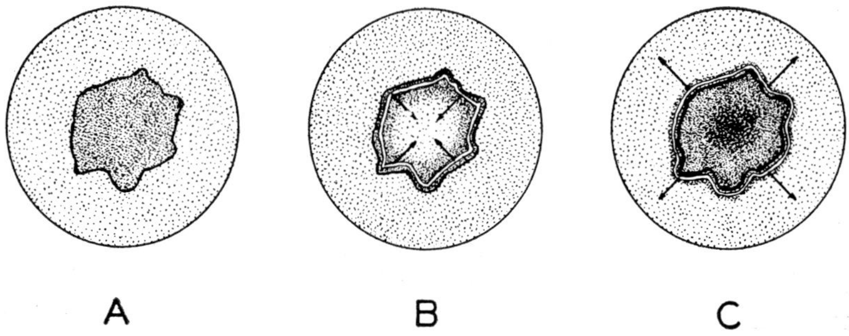

Fig. 6. Central illumination and the Becke line.

A. Microscope focused on grain.
B. Microscope tube slightly raised, index of fragment higher than that of surrounding medium.
C. Microscope tube slightly raised, index of fragment lower than that of surrounding medium.

of oblique illumination. In this method half the field of the microscope is darkened by the partial insertion of an accessory plate or some other opaque object into the optical system of the microscope so as to cut out half of the slightly converging cone of light that normally forms above the low-power substage condenser.

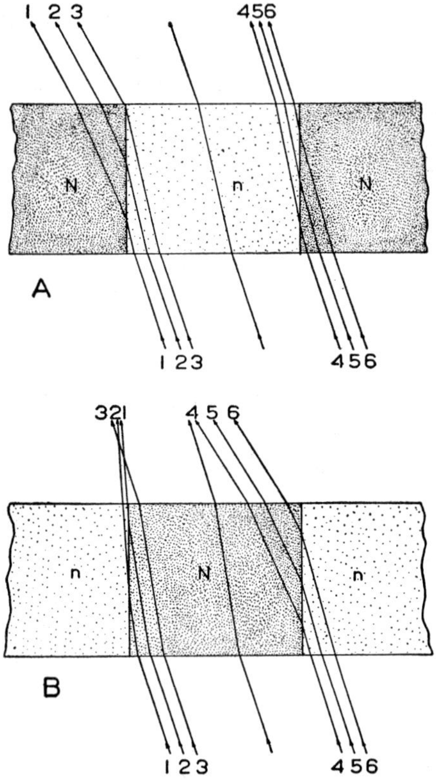

FIG. 7. Oblique illumination with vertical contacts. Reflected light not shown.

A. Index of substance *n* less than that of surrounding medium *N*.
B. Index of substance *N* greater than that of surrounding medium *n*.

Figures 7 and 8 illustrate crystal plates or fragments in obliquely incident light. In Fig. 7 the crystal makes a vertical contact with the immersion medium. In Fig. 7*A* it is supposed that the index *n* of the crystal is less than that of the immersion medium *N*. Light in inclined rays 1, 2, and 3 passing from the medium of index *n* into the medium of index *N* is refracted toward the normal and is dispersed. Light in rays 4, 5, and 6, passing from the medium of index *N* to the medium of index *n*, is refracted away from the normal and crowded

together (reflected light is not shown). If the fragment is viewed from above, one part appears dark, the other illuminated.

Figure 7*B* illustrates the refraction of light at vertical contacts when a substance has a higher index than that of the immersion medium. The result again will be the darkening of one part of the fragment and the illumination of the other, but the illuminated area is on the opposite side when compared to the example described above, in which the substance has a lower index than that of the immersion medium.

Lenticular fragments produce similar effects, as shown in Fig. 8. Figure 8*A* shows refraction of light by a lenticular fragment when its index is less than that of the immersion medium. Figure 8*B* illustrates the opposite condition.

The total result under the microscope, whether the substance has vertical sides or is lenticular, is the same. If half of the field is darkened, the fragments will be illuminated on one side and darkened on the other. The particular result depends on just where in the optical system the opaque object is inserted to darken half the field and on the particular combination of lenses used.

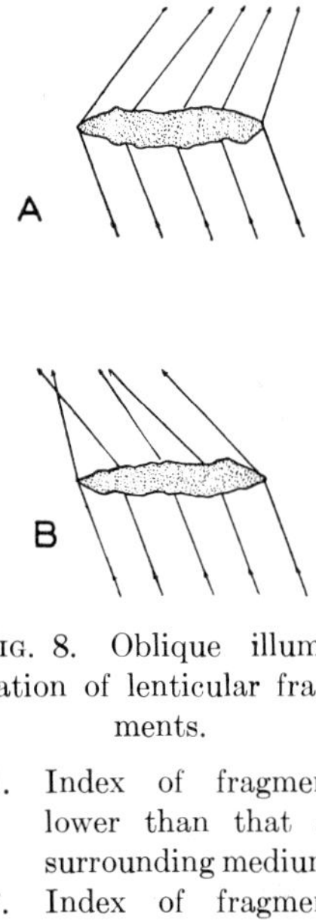

FIG. 8. Oblique illumination of lenticular fragments.

A. Index of fragment lower than that of surrounding medium.
B. Index of fragment higher than that of surrounding medium.

Best results are obtained with this method if the low-power objective is used and the field is half-darkened in the same manner for each measurement. It is advisable to determine the result in a fragment of known index immersed in a medium of known index. In subsequent determinations the identical arrangement of lenses should be preserved, and the field should be darkened in the same manner.

As in the method of central illumination, relief is a guide to the relative indices of the fragment and immersion medium. As the index of the fragment approaches that of the surrounding medium, the relief becomes less.

Figure 9 illustrates the effect of oblique illumination. In Fig. 9*A* the microscope is focused on two fragments immersed in oil. Figure 9*B* shows the effect of insertion of an accessory plate above a low-power objective when the substance has a lower index than the oil. The fragments have dark borders on the sides toward the dark half of the field and a bright border on the opposite side. In Fig. 9*C* the substance has a higher index than the oil, and the dark and light borders are reversed.

In monochromatic light, a transparent substance practically disappears when its index is exactly the same as that of the immersion medium.

Color Fringes in Oblique Illumination. Liquid immersion media generally disperse light more than the solids placed in them for comparison, and the difference between the indices for red and violet, at opposite ends of the spectrum, is greater in liquids than in most

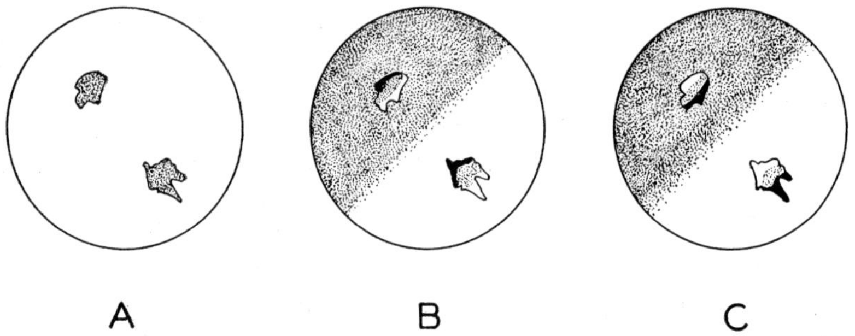

FIG. 9. Oblique illumination in plan.

A. Entire field illuminated.
B. Field half-darkened, index of fragments lower than that of oil. See text discussion.
C. Field half-darkened, index of fragments higher than that of oil. See text discussion.

solids. If, in the method of oblique illumination, white light is used and the index of a fragment for yellow light matches the index for yellow of the immersion liquid, color fringes appear around opposite edges of the fragment—red on one edge, blue (violet) on the other.

REFRACTIVE INDICES OF A CRYSTAL AND AN IMMERSION OIL AT 20° C

	Red 656mμ	*Yellow* 589mμ	*Violet* 434mμ
Crystal	1.519	1.520	1.521
Oil	1.516	1.520	1.524

The color fringes may be explained by reference to the accompanying table which gives the refractive indices corresponding to certain wave lengths for a hypothetical isotropic crystal in an immersion oil. At constant temperature the refractive index varies inversely as the wave length, and the difference between the indices for the extreme colors of the spectrum is greater in the oil than in the crystal. Suppose that the refractive indices for yellow light (589mμ) of crystal

and oil are matched. It then becomes apparent that at the red end of the spectrum the index for the oil is less than that of the crystal; at the other end of the spectrum the reverse is true.

Thus, in oblique illumination in white light, red and blue (violet) fringes appear when the indices of a fragment and the immersion medium are matched for a color in the middle of the spectrum. Unless otherwise stated most immersion media are standardized for a wave length of 589 millimicrons in the yellow portion of the spectrum. As other colors are matched in the crystal and the immersion medium the color fringes on either side of the fragment change accordingly. The oblique illumination method of determining refractive index in routine work yields results with an accuracy of ± 0.001 to ± 0.003.

Immersion Media. Immersion media are described by Larsen and Berman.[1] A brief discussion is presented here.

By means of immersion media it is possible to determine indices of refraction up to 3.17_{Li}. For accurate work media of low dispersion are desirable, but high dispersion aids in rapid determination. The indices of refraction of the media change with temperature but not at a constant rate. Accurate work should take into account index change as a function of temperature change. In general, the refractive index decreases as the temperature rises. The change of index with temperature change is relatively unimportant in solids.

Liquids may be used satisfactorily to determine indices between 1.33 and 2.06. Above 2.06 solids of low melting temperature may be used. Commonly used liquids and solids are listed in the accompanying table by Larsen and Berman. Liquid mixtures of sulfur, phosphorus, and methylene iodide with an index range of 1.74 to 2.06 are described by West.[2]

The refractive indices of immersion liquids within the range of 1.40 to 1.70 are conveniently measured with an Abbe refractometer. Liquids with an index up to 1.85 may be measured on the glass hemisphere of a refractometer of the Pulfrich type, but this instrument is not used widely because of its cost and delicacy. Liquids or low-melting-temperature solids may be measured conveniently and accurately in a hollow glass prism by the method of minimum deviation.

Refractive Index Dispersion. The index of refraction of an isotropic substance varies progressively with the wave length of the

[1] E. S. Larsen and Harry Berman, *The Microscopic Determination of the Nonopaque Minerals*, Bull. 848, U. S. Geological Survey, pp. 11–18, 1934.

[2] C. D. West, "Immersion liquids of high refractive index," *Am. Mineralogist*, Vol. 21, pp. 245–249, 1936.

REFRACTIVE INDICES OF IMMERSION MEDIA*

(From Larsen and Berman)

	n at 20° C	$-\frac{dn}{dt}$	Dispersion	Remarks
Water	1.333	Slight	....	Dissolves many of the minerals with low indices.
Acetone	1.357	....	Slight	Dissolves many of the minerals with low indices.
Ethyl alcohol†	1.362	0.00040	Slight	
Ethyl butyrate	1.381	....	Slight	
Methyl butyrate	1.386	....	Slight	
Ethyl valerate	1.393	....	Slight	
Amyl alcohol‡	1.409	0.00042	Slight	Dissolves many minerals with which it is used.
Kerosene	1.448	0.00035	Slight	
Petroleum oil§				
Russian alboline	1.470	0.0004	Slight	
American alboline	1.477	0.0004	Slight	
α-Monochlornaphthalene‖	1.626	....	Moderate	
α-Monobromnaphthalene	1.658	0.00048	Moderate	
Methylene iodide	1.737 to 1.741	0.00070	Rather strong	Rather expensive. Discolors on exposure to light, but a little copper or tin in the bottle will prevent this change.
Methylene iodide saturated with sulfur	1.778	....	Rather strong	
Methylene iodide, sulfur, and iodides#	1.868	....	Rather strong	
Piperine and iodides	1.68 to 2.10	....	....	
Sulfur and selenium	1.998_{Na} to 2.716_{Li}	....	Very strong	
Selenium and arsenic selenide	2.72 to 3.17_{Li}	....	Very strong	

* Another list of immersion media has been given by R. C. Emmons (*Am. Mineralogist*, Vol. 14, pp. 482–483, 1929).

† V. T. Harrington and M. S. Buerger used petroleum distillates prepared from kerosene, gasolene, etc., for the range of 1.35 to 1.45; the liquids with the lower indices are very volatile ("Immersion liquids of low refraction," *Am. Mineralogist*, Vol. 16, pp. 45–54, 1931).

‡ Ordinary fusel oil may be used, but on mixing with kerosene it forms a milky emulsion, which settles on standing, and then the clear liquid may be decanted off.

§ Any of the medicinal oils may be used, such as Nujol.

‖ Hallowax oil is satisfactory.

To 100 grams methylene iodide add 35 grams iodoform, 10 grams sulfur, 31 grams SnI_4, 16 grams AsI_3, and 8 grams SbI_3, warm to hasten solution, allow to stand, and filter off undissolved solids. (See H. E. Merwin, "Media of high refraction," *Washington Acad. Sci. Jour.*, Vol. 3, pp. 35–40, 1913.)

transmitted light; also, the principal indices of refraction of birefringent substances vary individually in the same manner. The index variation is an expression of the fact that for a single angle of incidence the angles of refraction differ for the different wave lengths of light. This inequality of refraction of the progressively varying wave lengths of light differs in amount in different media. The degree of inequality of refraction is a measure of the *dispersion.*

Certain well-spaced and easily obtained monochromatic radiations have been used as standards for specifying refractive index in terms of the wave length of light. Many of these wave lengths correspond to certain lines in the sun's spectrum and have been called *Fraunhofer lines.* The accompanying table gives the letter designations, wave lengths in millimicrons, and common sources for many of the standard wave lengths.

REFERENCE WAVE LENGTHS IN THE VISIBLE SPECTRUM

Letter Designation	*Wave Length* mμ	*Source*
A^1	766.5	Potassium flame
C	656.3	Hydrogen discharge
D	589.3	Sodium flame
d	587.6	Helium discharge
e	546.1	Mercury arc
E	527.0	Sun
F	486.1	Hydrogen discharge
g	435.9	Mercury arc
G^1	434.1	Hydrogen discharge
h	404.7	Mercury arc

The *total dispersion* may be expressed as the difference between the refractive indices for red and violet light. *Partial dispersion* measures the difference between the indices for specified wave lengths within the spectrum.

Relative dispersion or *dispersive power* δ is given by the following expression:

$$\delta = \frac{n_F - n_C}{n_D - 1}$$

where n_F, n_C, and n_D are the refractive indices for the Fraunhofer lines F, C, and D.

Relative dispersions and their reciprocals are useful in computations for the construction of achromatic lenses.

A simple relationship between refractive index and wave length is

given by Cauchy's equation, which follows:

$$n = A + \frac{B}{\lambda^2} + \frac{C}{\lambda^4} \cdots$$

where n is the refractive index, λ is the wave length, and A, B, and C are constants.

Now, if the refractive indices of a substance for three different wave lengths are determined, and each of the values is substituted in the above equation, three simultaneous equations which permit evaluation of A, B, and C are obtained. When these constants are determined for a given substance, it becomes possible to compute the index for any wave length of the visible spectrum.

Hartmann equations show an empirical relationship between refractive index and wave length, as follows:

$$n = n_0 + \frac{c}{(\lambda_0 - \lambda)^{1.2}}$$

and

$$n = n_0 + \frac{c}{\lambda - \lambda_0}$$

Refractive index is a function not only of wave length but of temperature. In general, refractive indices decrease as temperature is elevated. The indices of liquids change much more rapidly with temperature than the indices of solids. Accordingly, both temperature and the wave length of light should be specified in stating the results of precise index measurement.

Figure 10 shows the variation of refractive index with temperature and wave length for ethyl salicylate, a liquid, and the variation of the index with wave length for a borosilicate glass. The curves in Fig. 10 may be described as *dispersion curves.* Data plotted on a graph with a uniform wave length scale yield curved lines, but the same data plotted on diagrams with a logarithmic wave length scale as in Fig. 10 yield essentially straight lines that permit fairly accurate interpolation and extrapolation.

In Fig. 10 note the wide variation in the refractive indices for the liquid with wave length and temperature. The index of the glass changes appreciably with wave length, but index changes corresponding to temperature changes are so small that for small temperature intervals they are practically negligible. If it is assumed that the refractive index of the glass does not change appreciably in the temperature range between 10° and 50° C, it can be seen that the

index of the liquid at either 10° or 50° C can be made to match the index of the glass by changing the wave length of the light. At 10° C liquid and glass have the same refractive indices for a wave

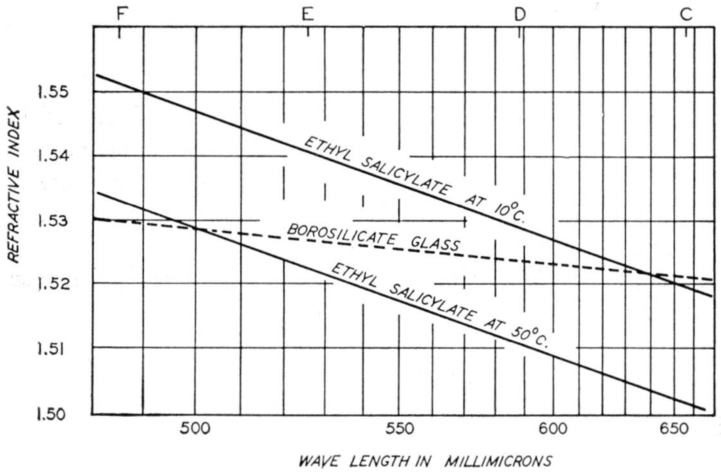

FIG. 10. Dispersion curves showing relationship between index of refraction and wave length and temperature.

length of 640 millimicrons; at 50° C the indices are the same at a wave length of 500 millimicrons.

Dispersion Methods of Index Measurement. Merwin[1] has described a procedure for the accurate determination of index of refraction by the embedding method. A grain of the substance to be measured is embedded in a liquid of a slightly higher index for sodium light. The grain is examined in monochromatic light produced by a monochromatic illuminator, and the wave length is changed until the indices of the grain and the oil match. The grain is then embedded in an oil of slightly lower index for sodium light, and the wave length is again varied until the indices are matched. Inasmuch as the curve showing variation of index with the wave length of light as plotted on a logarithmic scale is nearly a straight line (the dispersion curve) for most solids, the index of a substance for any wave length of light is easily determined by interpolation or extrapolation. The indices of immersion oils for the various wave lengths of light may be determined from their dispersion curves or may be measured with a refractometer at the time the grain is measured.

[1] E. Posnjak and H. E. Merwin, "The system Fe_2O_3-SO_3-H_2O," *Am. Chem. Soc. Jour.*, Vol. 44, p. 1970, 1922.

Another method, Emmons'[1] *double-variation method,* utilizes the principle illustrated in Fig. 10 and measures index of refraction by embedding crystal fragments in liquids and varying both the wave length of light and the temperature of the liquid. A single mount may yield sufficient information to permit determination of all indices of refraction, each for any wave length of light.

Method of Minimum Deviation. One of the most accurate methods for measuring the index of refraction is the *method of minimum deviation.* This method requires crystals with faces intersecting at an angle somewhat less than 90 degrees or a prism cut from the substance to be measured. Use is made of a spectrometer or a one-circle reflecting goniometer. The prism to be measured is mounted, and the angle α between the faces is determined. A monochromatic light beam is then passed through the prism, the prism is rotated, and the refracted beam of light is followed with the telescope until it reaches the position of minimum deviation. The angle between the light in the rays in this position and the undeviated light from the source is the angle of minimum deviation δ.

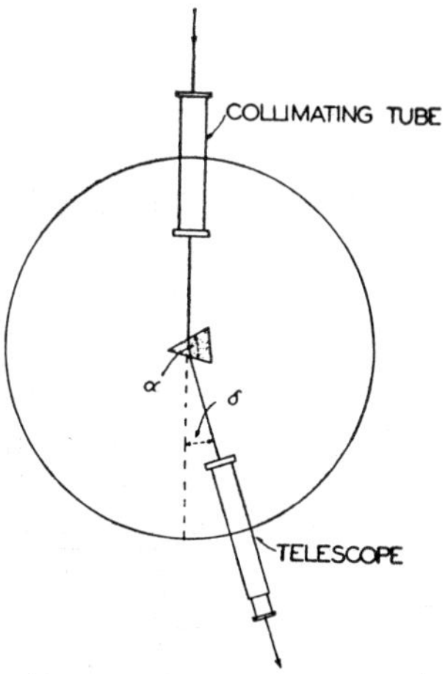

FIG. 11. Method of minimum deviation. Necessary measurements with goniometer.

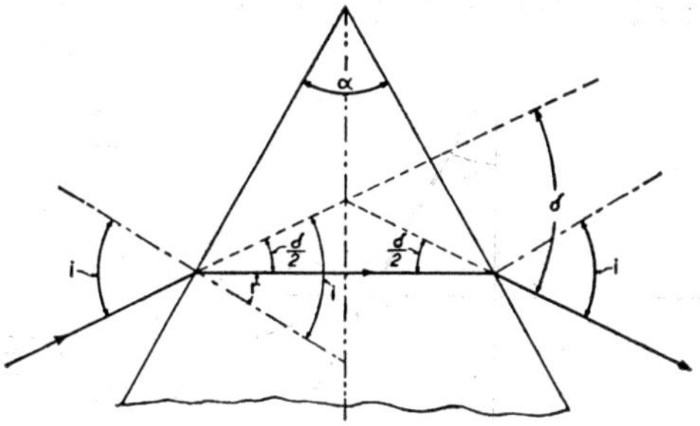

FIG. 12. Method of minimum deviation. Path followed by light at position of minimum deviation.

Figure 11 indicates diagrammatically the measurements required in the method of minimum deviation. Figure 12 shows in detail the passage of a light through a prism in the position of minimum deviation. Note that the angle of incidence of the entering light is the same as the angle of refraction of the emergent light. Now, let n be

[1] R. C. Emmons, "The double variation method of refractive index determination," *Am. Mineralogist,* Vol. 14, pp. 414–426, 441–461, 1929; and *The Universal Stage,* Mem. 8, Geol. Soc. Amer., 1943.

the index of refraction of the prism; then

$$n = \frac{\sin i}{\sin r}$$

$$r = \tfrac{1}{2}\alpha \text{ (mutually perpendicular sides)}$$

and

$$i = r + \tfrac{1}{2}\delta$$

Therefore

$$i = \tfrac{1}{2}\alpha + \tfrac{1}{2}\delta$$

Substituting

$$n = \frac{\sin i}{\sin r} = \frac{\sin \frac{1}{2}(\alpha + \delta)}{\sin \frac{1}{2}\alpha}$$

Method of Perpendicular Incidence. This method is not as widely used as the method of minimum deviation but employs the same instrument and gives equally accurate results. As in the method of minimum deviation, the angle between two faces of a crystal or a cut and polished prism is used. The angle α (Fig. 13) is measured, and the prism is rotated so that one of its faces is perpendicular to the beam of light from the collimating tube. The angular position of the refracted light passing through the telescope is determined, and the angle δ between the refracted light and the undeviated light is measured. Now

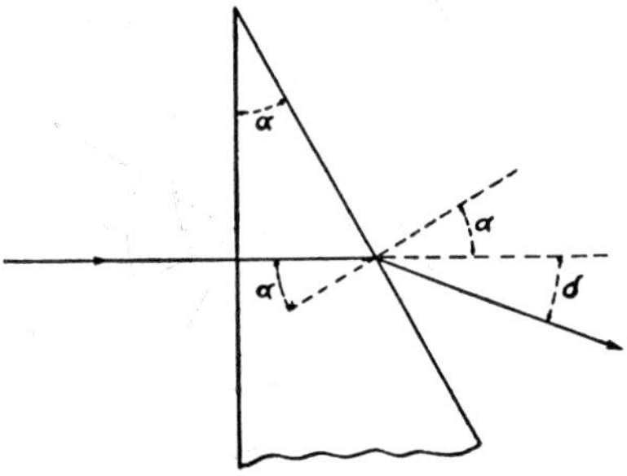

FIG. 13. Method of perpendicular incidence. Necessary measurements.

$$i = \alpha + \delta$$

and

$$r = \alpha$$

By substituting in the equation,

$$n = \frac{\sin i}{\sin r}$$

we obtain

$$n = \frac{\sin (\alpha + \delta)}{\sin \alpha}$$

It should be remembered that the angle i is the angle between the normal and the path followed by the light in air so that, although the

light is refracted as it passes from the prism into air, the angle of refraction at emergence is properly designated as i.

Methods Based on Critical Angle. There are many variations of the method based on measurement of the critical angle. A few will be described here to demonstrate the principles.

The Abbe total reflectometer utilizes glass hemispheres of high index of refraction (commonly about 1.85). A flat surface of the substance of unknown index is placed on the flat surface of the hemisphere after a drop of high-index liquid has been used to remove the film of air at the contact. If this method is to be successful, the index of the unknown substance must be less than that of the hemisphere.

If a source of slightly converging monochromatic light is placed below and to one side of the hemisphere, the light will pass through

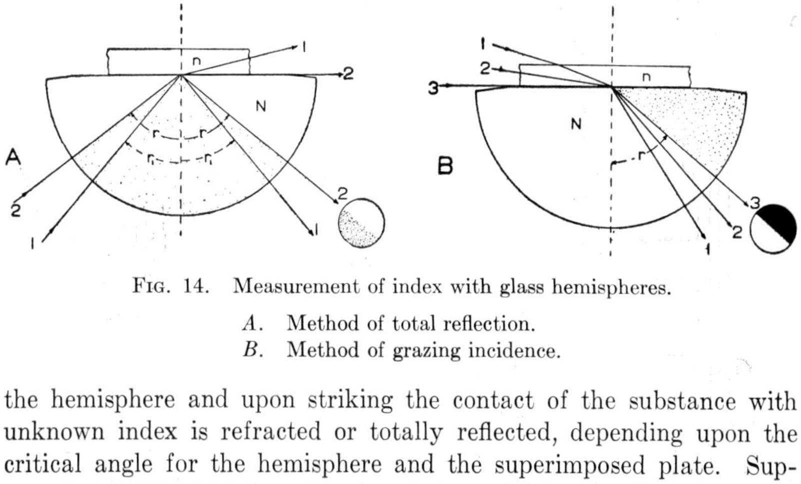

FIG. 14. Measurement of index with glass hemispheres.
A. Method of total reflection.
B. Method of grazing incidence.

the hemisphere and upon striking the contact of the substance with unknown index is refracted or totally reflected, depending upon the critical angle for the hemisphere and the superimposed plate. Suppose that in Fig. 14*A* the critical angle is r. Light traveling along rays of angle r_1, less than r, is partly refracted and partly reflected. Light in rays of greater angle is completely reflected. The totally reflected light is more intense than that which suffers partial refraction, and a telescope lined up with the reflected light in the critical position reveals a sharp contact between a bright field and one of diminished intensity. The angle r between the reflected light in the critical position and the normal to the flat surface of the hemisphere may be measured. The index of the hemisphere is already known, so that when r is measured, the index of the unknown substance may be computed.

If a monochromatic light source is placed above the hemisphere as in Fig. 14*B*, light passing the substance of smaller index n is refracted

toward the normal upon entering the hemisphere. Inasmuch as the maximum angle of incidence is 90 degrees (ray 3), no light will pass

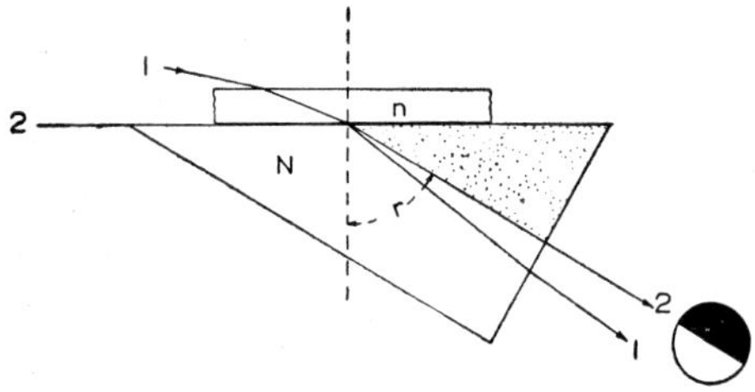

FIG. 15. Index determination with glass prism.

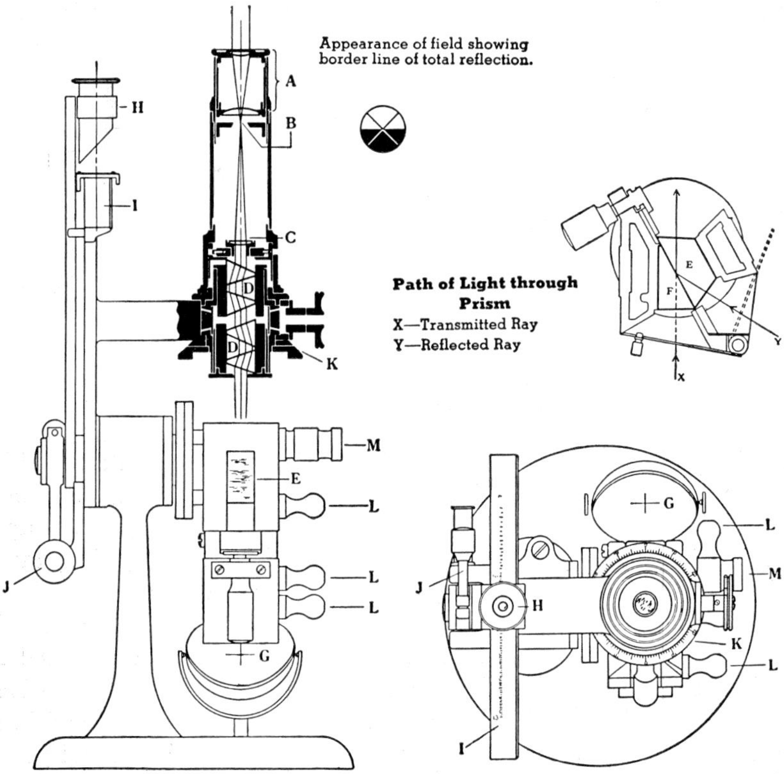

Courtesy of Bausch and Lomb Optical Co.

FIG. 16. Construction of Abbe refractometer.

through the hemisphere with an angle of refraction r greater than that of the light with an angle of incidence of 90 degrees. Accordingly, a telescope lined up with ray 3 will reveal a field which is half

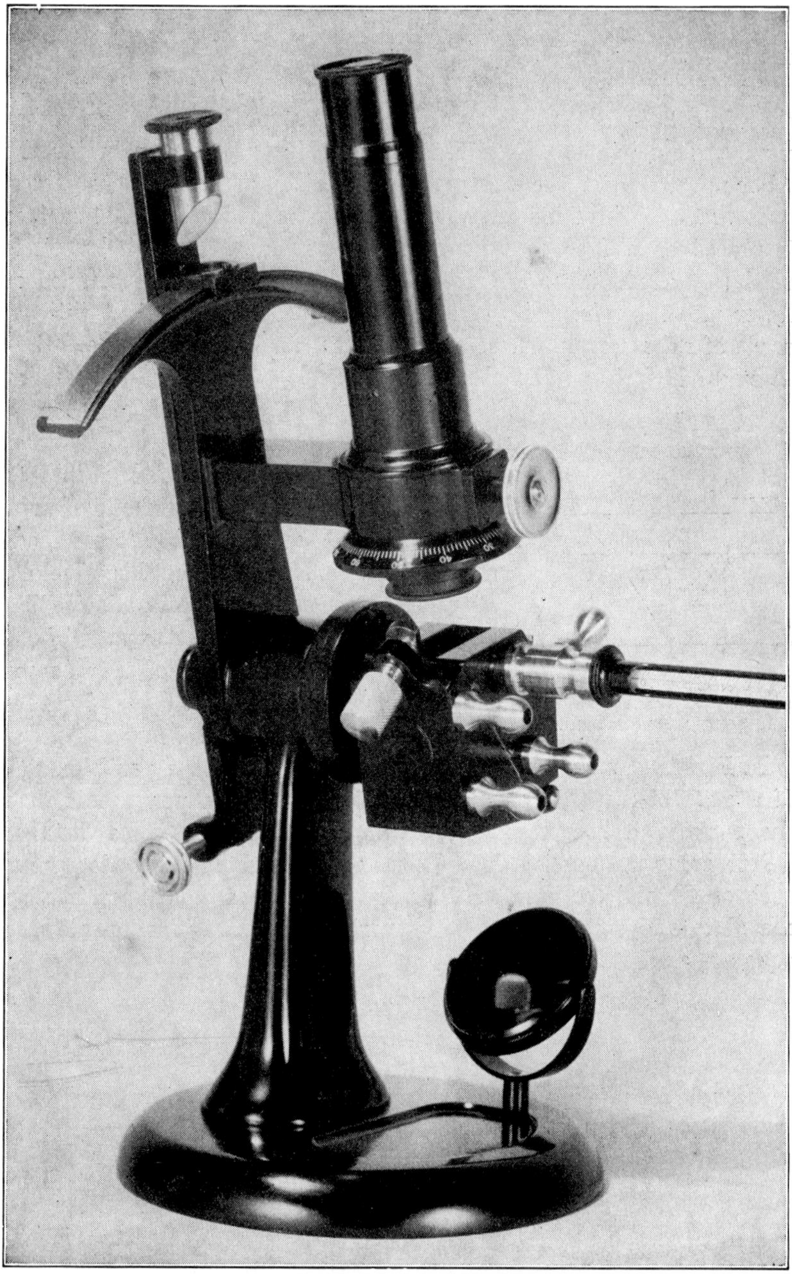

Courtesy of Bausch and Lomb Optical Co.

FIG. 17. Abbe refractometer.

dark and half illuminated. The angle r, as measured, yields the index of the superimposed plate by computation.

Another method which utilizes the principles of total reflection and the critical angle employs a prism or a combination of prisms rather than a glass hemisphere. Many modern instruments, including the Abbe refractometer, are constructed with prisms instead of hemispheres.

In practice, if the index n of a solid is to be measured, a flat surface is placed on a face of a prism of higher index. High-index oil is used to exclude the air film. Light passing from the substance of lower index is refracted when entering the prism. No light with an angle of refraction in excess of the critical angle r (Fig. 15) can pass through the prism. Inasmuch as the angle r varies with the index of the superimposed plate, it is possible to construct an instrument which, by means of a movable telescope and a fixed calibrated scale, gives the index of the superimposed substance directly. Figures 16 and 17 show a modern Abbe refractometer utilizing this principle.

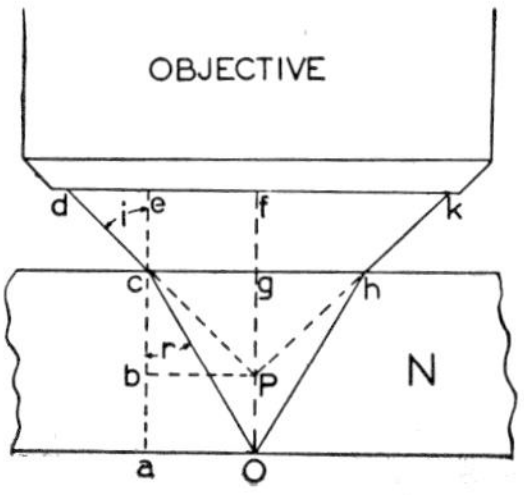

FIG. 18. Chaulnes' method of index determination.

Chaulnes' Method. The method devised by Duc de Chaulnes for the measurement of the index of refraction of plates of transparent substances is used occasionally and yields moderately accurate results.

In Fig. 18 suppose that the objective is focused on a particle at O. If a plate of transparent substance of index N is superimposed over the particle, it becomes necessary to raise the objective a distance OP to bring the particle into focus again. Light moving along rays such as Oc and Oh is refracted away from the normal upon leaving the plate and entering air. Now angle ecd equals i, and angle Oca equals r, and

$$N = \frac{\sin i}{\sin r}$$

In triangle cbP, angle bcP is equal to i, and

$$\tan i = \frac{bP}{cb}$$

In triangle caO angle acO equals r, and

$$\tan r = \frac{aO}{ca}$$

Now $aO = bP$, and inasmuch as the ratio of the tangents of small angles is essentially the same as the ratio of the sines,

$$N = \frac{\sin i}{\sin r} = \frac{\tan i}{\tan r} = \frac{bP/cb}{bP/ca} = \frac{ca}{cb}$$

Since $Og = ca$, and $gP = cb$,

$$N = \frac{Og}{gP}$$

Og and gP may be measured with the micrometer scale on the fine-adjustment screw of the microscope.

Interrelations of Index of Refraction, Specific Gravity, and Chemical Composition. The law of Gladstone and Dale states: "Every liquid has a specific refractive energy composed of the specific refractive energies of its component elements, modified by the manner of combination, and which is unaffected by changes of temperature and accompanies it when mixed with other liquids. The product of this specific refractive energy and the density is, when added to unity, the refractive index."

Mathematically, this law is expressed as follows:

$$K = \frac{n - 1}{d}$$

where K is the specific refractive energy of a substance, n is its index of refraction, and d is its density. Put another way, the law states that

$$K = k_1 \cdot \frac{p_1}{100} + k_2 \cdot \frac{p_2}{100} + \cdots$$

where k_1, k_2, etc., are the specific refractive energies of the components of the substance, and p_1, p_2, etc., are the weight percentages of the components of the substance. Gladstone and Dale's law under certain conditions yields poor results.

Lorentz and Lorenz independently derived a formula which obviates some of the difficulties inherent in Gladstone and Dale's law. Their formula states that

$$K = \frac{n^2 - 1}{n^2 + 2} \cdot \frac{1}{d}$$

Lichtenecker proposed a further modification, which is

$$K = \frac{\log n}{d}$$

A serious defect with all the proposed equations is that in some substances the specific refractive energy of one or more components depends upon the structure of the molecule.

Gladstone and Dale's law may be used to calculate approximately the average index of a crystalline solid if its density and the specific refractive energies of its components are known. Larsen and Berman[1] tabulate the specific refractive energies of the constituents of minerals, and they discuss the use of specific refractive energies in detail.

The *molecular refraction* of a substance may be computed from the following equation

$$N = \frac{M(n^2 - 1)}{d\,(n^2 + 2)}$$

where N is the molecular refraction for a specified temperature and wave length, M is the molecular weight, d is the density, and n is the refractive index.

[1] E. S. Larsen and Harry Berman, *The Microscopic Determination of the Nonopaque Minerals*, Bull. 848, U. S. Geological Survey, pp. 11–18, 1934.

CHAPTER VII

THE UNIAXIAL INDICATRIX

Introduction. Monochromatic light travels with equal velocity in all directions in isotropic substances; moreover, the light, except insofar as it is partially polarized by refraction and reflection, is not constrained to vibrate in any particular direction. This is not true in uniaxial crystals, for in them the light is in general polarized so as to vibrate in two mutually perpendicular planes. Light transmitted along each ray or wave normal within the crystal has a velocity which depends on its direction of propagation or vibration.

There is one direction in uniaxial crystals along which all monochromatic light moves with the same velocity. This direction, which is parallel to the c crystallographic axis, is called the optic axis, and, inasmuch as there is only *one* such direction, the crystals are described as *uniaxial.*

For each wave length of light uniaxial crystals have *two principal indices of refraction,* from which it follows that light traveling in any direction except the optic axis consists of two components with different velocities. The change of refractive index with the direction of light propagation or vibration may be visualized in a *uniaxial indicatrix,* a three-dimensional geometric figure showing the variation of the indices of refraction of light waves in their *directions of vibration.* Each radius vector represents a vibration direction whose length measures the index of refraction of a wave vibrating parallel to it.

Geometric Representation of Index Variation—The Indicatrix. Figure 1 shows indicatrices for positive and negative uniaxial crystals. Figure 1*A* is a prolate spheroid of rotation constructed so that its semimajor and semiminor axes are proportional, respectively, to the maximum and minimum refractive indices of a uniaxial positive mineral. Figure 1*B* portrays a negative uniaxial indicatrix: an oblate spheroid of rotation. Any section passing through and including the optic axis of either indicatrix is an ellipse and is called a *principal section.* Equatorial sections at right angles to the optic axis are circular in outline.

Monochromatic light traveling along the optic axis in either Fig. 1*A* or 1*B* vibrates in the same manner at right angles to the optic axis in

all directions parallel to the radii of the circular equatorial section. In the direction of the optic axis, and in this direction only, all light travels through the crystal as if it were isotropic. Moreover, monochromatic light traveling in any direction whatsoever through the crystal and vibrating parallel to a radius of the equatorial section has the same refractive index and the same velocity. Accordingly, waves which vibrate parallel to the radii of the circular equatorial section (perpendicular to the optic axis) are referred to as *ordinary* or *O waves;* the refractive index of these waves is described as the *index*

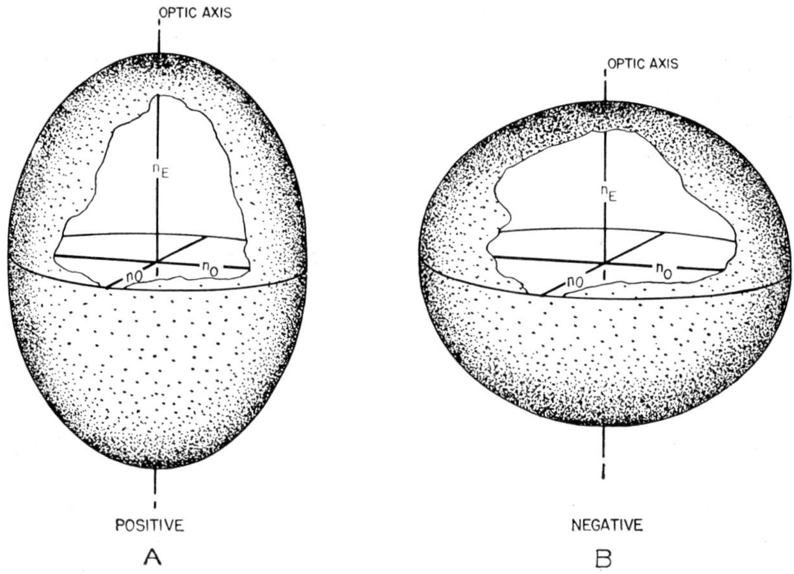

FIG. 1. Positive and negative uniaxial indicatrices.

of the ordinary wave, and is here designated as n_O. Rays which transmit light vibrating parallel to the radii of the equatorial section are identified as *ordinary* or *O rays.*

Waves vibrating in a plane including the optic axis (a principal section) have refractive indices and velocities which depend on their direction of vibration or propagation and are called *extraordinary* or *E waves.* The index of a wave vibrating parallel to the optic axis in a principal section is at a maximum or minimum, depending upon whether the crystal is positive or negative. The index of such a wave is designated as n_E. A wave vibrating in a principal section and traveling in a random direction has a refractive index somewhere between n_O and n_E, the exact index being a function of the direction of

vibration and the configuration of the indicatrix. Rays which transmit light vibrating in a principal section are described as *extraordinary* or *E rays*.

The indicatrix gives the values for the refractive indices of all waves in their direction of vibration for only *one* wave length of light. For each wave length of the spectrum there is, for each substance, a characteristic indicatrix. In general, n_O and n_E increase as the wave length decreases, but not at the same rate.

The use of n_O and n_E to designate the principal refractive indices of uniaxial crystals may cause some confusion. In the 1943 edition of this book ω and ϵ were used, but in order to introduce a certain element of consistency with modern usage, it was decided to make the indicated change. No confusion should result if the equivalent usage shown in the accompanying table is kept in mind.

VARIOUS DESIGNATIONS OF PRINCIPAL REFRACTIVE INDICES OF UNIAXIAL CRYSTALS

Index of Ordinary Wave	n_O	N_O	O	ω	n_ω	N_ω
Principal Index of Extraordinary Wave	n_E	N_E	E	ϵ	n_ϵ	N_ϵ

From Figs. 1*A* and 1*B* it is seen that in positive uniaxial crystals $n_E > n_O$, and in negative crystals $n_E < n_O$. As n_E approaches n_O, the indicatrix approaches the form of a sphere, and when n_E finally equals n_O, the crystal becomes isotropic.

An important use of the indicatrix is suggested in Fig. 2, which shows a principal section of a uniaxial positive indicatrix with $n_E = 2.0$ and $n_O = 1.5$. O is a point source of monochromatic light; O' is a second point source and is shown for the purpose of emphasizing the fact that in constructions involving the indicatrix, it is the *direction* of vibration or propagation of light, *not* the specific path followed by a light wave as related to a single indicatrix that is of most concern.

Suppose that a wave of light moves in the direction of its wave normal ON'. The wave vibrates parallel to TQ, which is perpendicular to ON', and by construction has a refractive index equal to OT or OQ. In the example shown, the index of refraction of the wave is 1.68. Another wave moving along ON' vibrates normal to the plane of the drawing and has an index of 1.5, equal to the radius of the circular equatorial section of the indicatrix.

In general, the direction of movement of a wave does not coincide with the direction of propagation of light along the rays. Moreover, it is believed, from consideration of electromagnetic-wave theory, that light moving along a ray vibrates in a direction perpendicular to

the associated wave normal. Accordingly, the vibration direction of light transmitted along a ray need not be at right angles to the ray path. Exceptions to this rule occur when the rays are parallel to the optic axis or to an equatorial radius of the indicatrix. In these directions the rays are parallel to their associated wave normals.

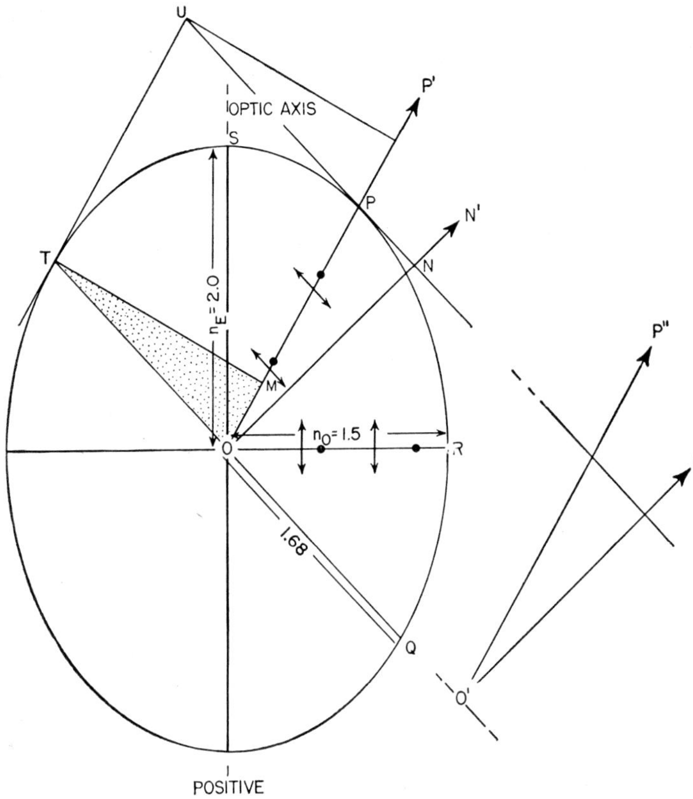

FIG. 2. Principal section of positive uniaxial indicatrix showing relationships of rays and waves.

To obtain the direction of the rays corresponding to a wave moving in the direction ON', a tangent TU is drawn to the ellipse at T. The line OP', parallel to TU, gives the direction of the rays and another tangent to the ellipse at P, the point of intersection of the ray and the indicatrix, is parallel to the wave front. The radii OT and OP are *conjugate radii;* they fulfill the specification that one radius be conjugate to a second radius if the first radius is parallel to the tangent to the ellipse at the end of the second radius.

The refractive index of the wave moving along ON' is OT (or OQ), but light following the ray OP' does not move in the same direction as the wave and must have a different refractive index. The refractive index of the light moving along the ray is obtained by dropping a line TM from T perpendicular to OP'. TM can be thought of as the effective cross section of OT, the index of the wave traveling in the direction ON', giving the index of the light traveling along the ray OP'. Thus, OT properly may be designated as the *wave index of refraction,* and TM as the corresponding *ray index of refraction.* The ray index refers to the light traveling along the ray as contrasted with the light traveling in the direction of the wave normal.

OT is greater than TM, and, inasmuch as the velocities of light moving in the direction of the wave normal and light moving along a ray are inversely proportional to their respective refractive indices, it is seen that the velocity of the wave moving along ON' is less than the velocity of the light propagated along the ray OP'.

Light moving along OP' and vibrating perpendicular to the plane of the drawing in Fig. 2 travels with a velocity proportional to $1/OR = 1/n_O = 1/1.5$.

All light moving along OS, the optic axis, has a constant refractive index and vibrates at right angles to the optic axis. In Fig. 2 this light has a velocity proportional to $1/n_O = 1/1.5$.

Light moving along OR consists of two components vibrating in mutually perpendicular planes. The component vibrating in the principal section has an index of n_E, and the component vibrating perpendicular to the principal plane has an index of n_O. Along OR light in the rays vibrates and travels in the same direction as the waves, and the ray and wave indices of refraction for each component are equal.

Ray Velocity Surfaces. Ray velocity surfaces (ray surfaces) are geometrically and mathematically related to the indicatrix and permit visualization of the velocities of propagation of light along its rays in all directions in a crystal.

Figure 3 portrays the relationship between the indicatrix and the ray velocity surfaces in a negative crystal. Light following a ray OP consists of two components, one vibrating in the principal section (plane of drawing) and the other perpendicular to the principal section. In negative crystals, the component vibrating in the principal section is faster than the component vibrating normal to it, as is shown by the spacing of the dots and arrows in the diagram. The ordinary component of light traveling along ray OP_1, for example, in a given instant will move a distance proportional to $1/n_O$. Along the

other paths OP_2, OP_3, etc., the ordinary component moves the same distance in the same instant. In two dimensions the velocity of the ordinary component of light moving along all the rays is represented

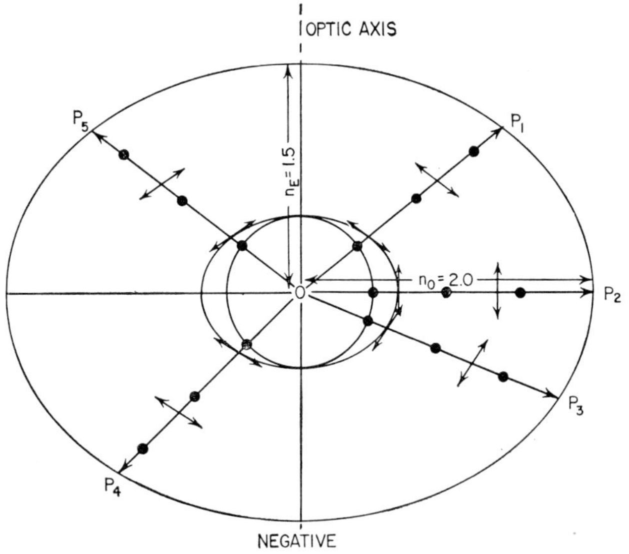

FIG. 3. Principal section of negative uniaxial indicatrix showing relationship of ray velocity surfaces to indicatrix.

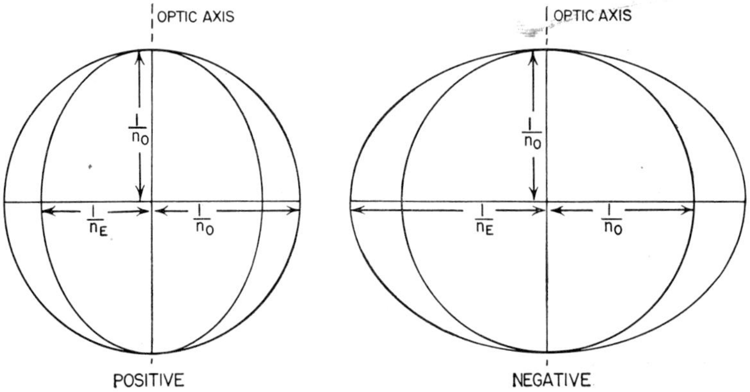

FIG. 4. Principal sections of positive and negative ray velocity surfaces.

by a circle with its center at O. In three dimensions the velocity is represented by a sphere with its center at O.

The velocity of the extraordinary component of light traveling along a ray is inversely proportional to its ray index of refraction. The

extraordinary component, in negative crystals, vibrates in the principal section and moves with maximum velocity in a direction perpendicular to the optic axis (velocity proportional to $1/n_E$). The extraordinary component has a minimum velocity equal to the velocity of light traveling along the ordinary ray when it travels parallel to the optic axis. This relationship is reversed in positive crystals. The variation in the velocities of components moving along the extraordinary rays is represented in cross section by an ellipse and in three dimensions by an ellipsoid of rotation.

Figure 4 shows principal sections of ray velocity surfaces

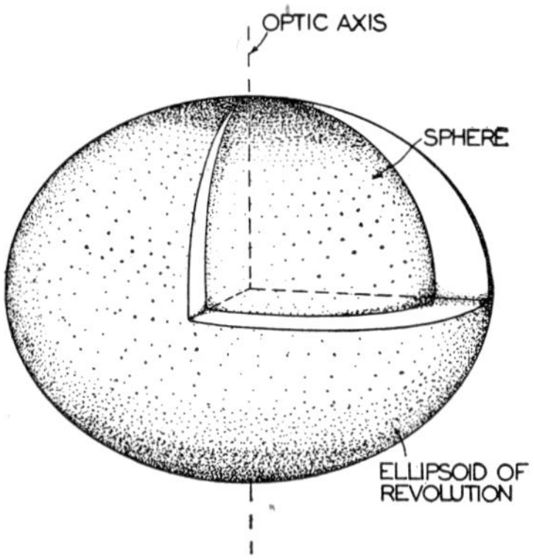

FIG. 5. Uniaxial negative ray velocity surface sectioned to show sphere within an ellipsoid of rotation.

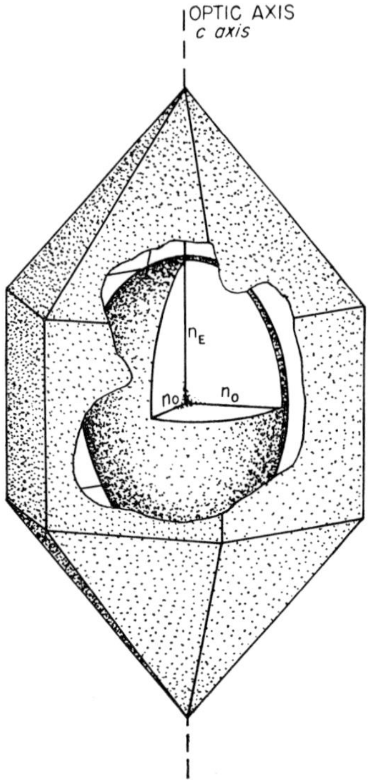

FIG. 6. Uniaxial positive crystal (quartz) showing orientation of the indicatrix. Rotary polarization disregarded.

of positive and negative crystals. Sections drawn normal to the optic axis consist of two concentric circles.

To assist in the visualization of the ray velocity surfaces in three dimensions, Fig. 5 is useful. The drawing depicts the ray velocity surface for a negative crystal sectioned so as to show the spherical surface for the ordinary rays enclosed by an oblate ellipsoid of rotation representing the velocities of light in the extraordinary rays.

Positive ray velocity surfaces in three dimensions consist of a prolate spheroid of rotation enclosed by a sphere.

Optical Orientation of Uniaxial Crystals. Symmetry requirements of tetragonal and hexagonal crystals are such that the optic axis must parallel the *c* crystallographic axis. Any other orientation would result in disharmony of the symmetry elements of indicatrix and crystal.

Figures 6 and 7 show the orientations of the indicatrices in quartz and calcite. Visualization of these relationships is very useful in studying the optics of uniaxial crystals, for similar conditions exist in all uniaxial crystals.

The direction of the *c* axis of a crystal or fragment under the microscope is determined when the vibration direction corresponding to the

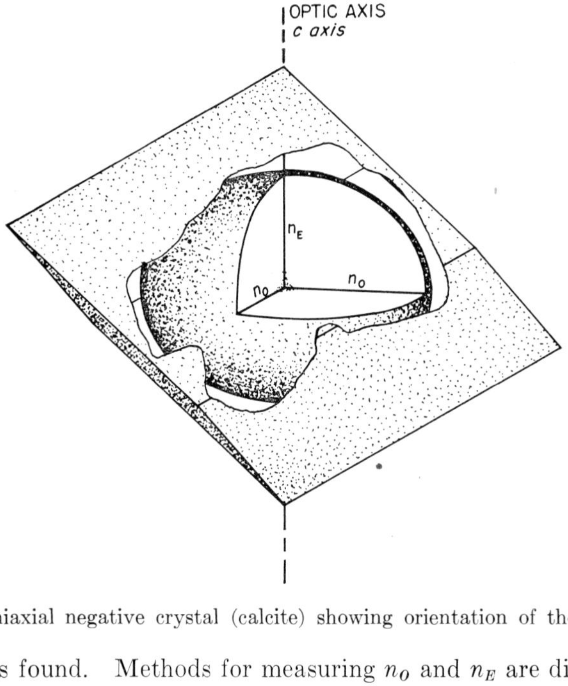

FIG. 7. Uniaxial negative crystal (calcite) showing orientation of the indicatrix.

n_E index is found. Methods for measuring n_O and n_E are discussed in a subsequent chapter.

Huygenian Constructions. Ray velocity surfaces serve one of their most useful functions in the geometric study of the passage of light through crystals. By using Huygenian constructions it is possible to trace the path of refracted polarized light through a crystal for light of any angle of incidence. Huygenian constructions are usually drawn in planes of crystal symmetry, because in these planes the paths followed by the light may be represented graphically in two dimensions. Huygenian constructions not in a plane of symmetry require difficultly constructed three-dimensional drawings.

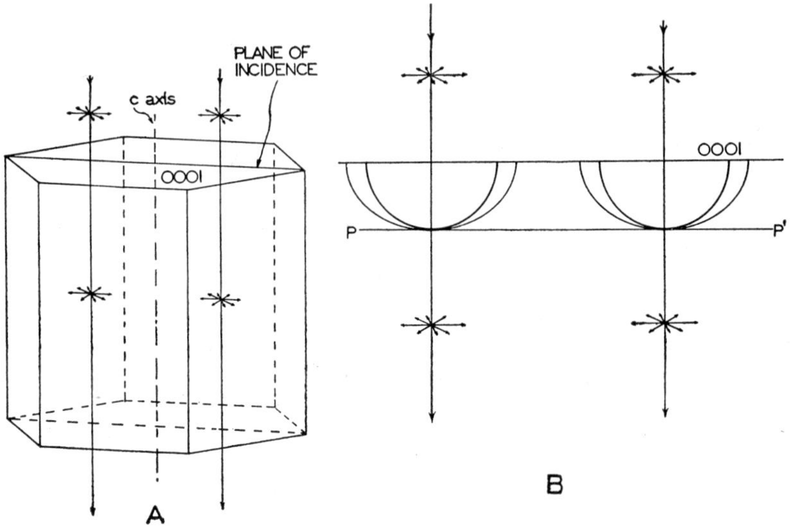

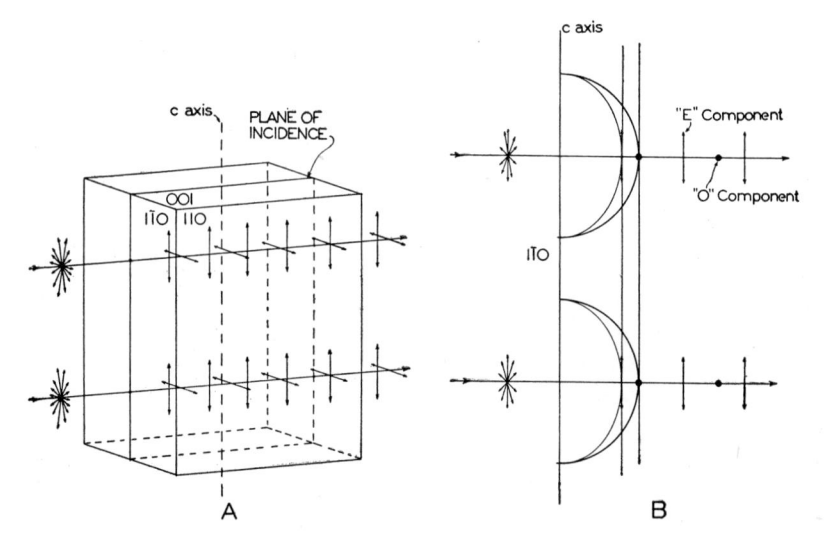

FIG. 8. Light perpendicularly incident on base of a negative hexagonal crystal.

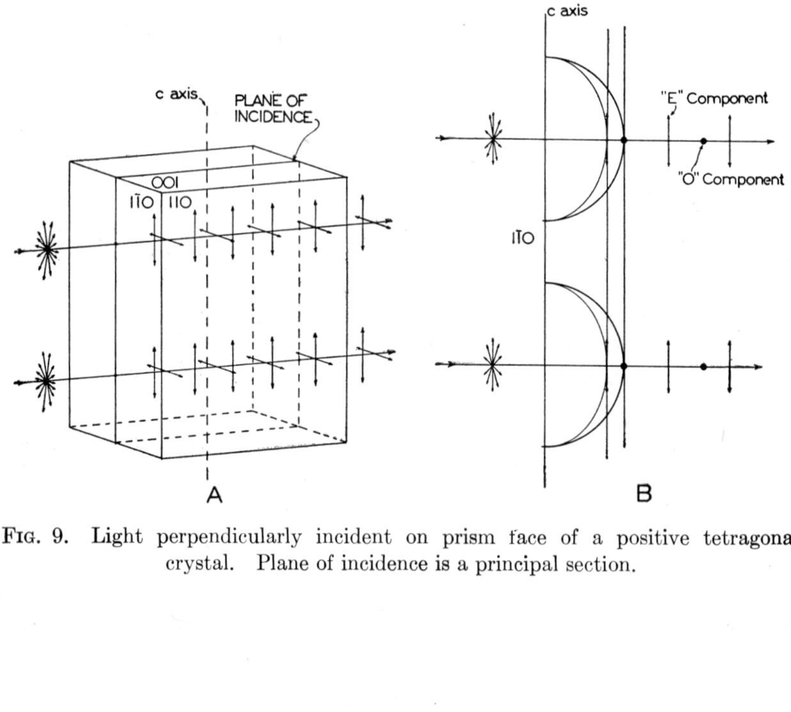

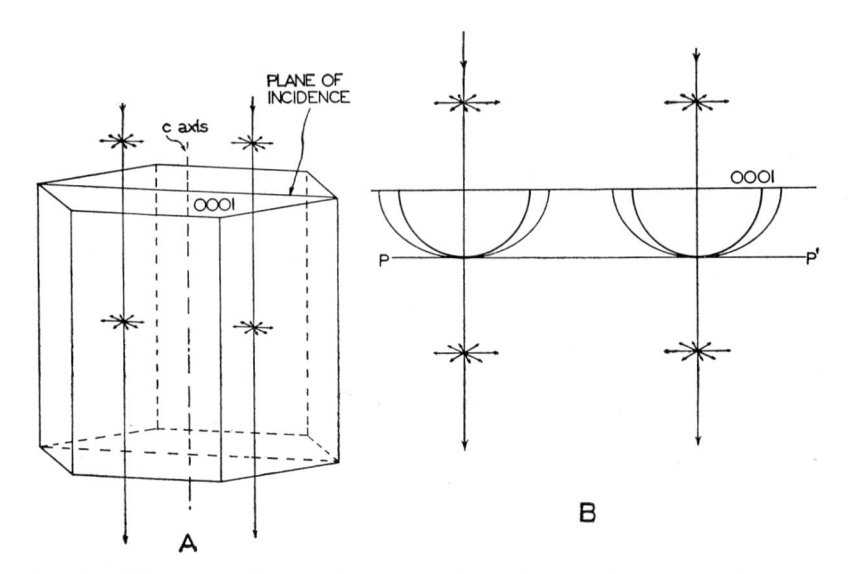

FIG. 9. Light perpendicularly incident on prism face of a positive tetragonal crystal. Plane of incidence is a principal section.

Figure 8 shows light perpendicularly incident on the basal plane of a negative hexagonal crystal. The light is not polarized and passes through the crystal as if the crystal were isotropic. The reason for this becomes apparent upon examination of Fig. 8*B*, which lies in a

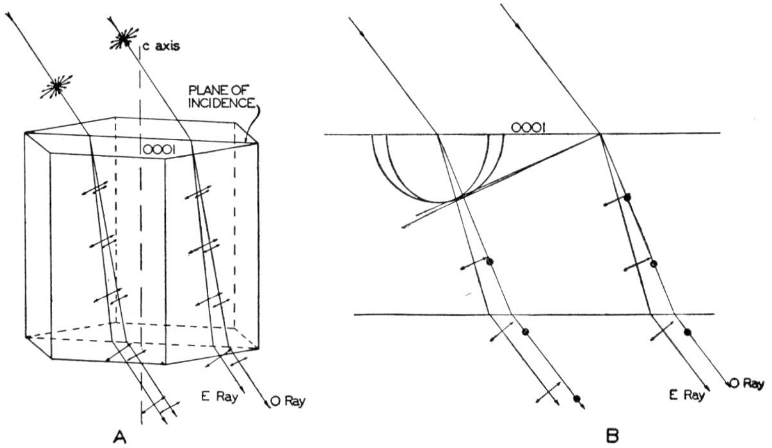

FIG. 10. Light falling with inclined incidence on base of a positive hexagonal crystal. Plane of incidence is a principal section.

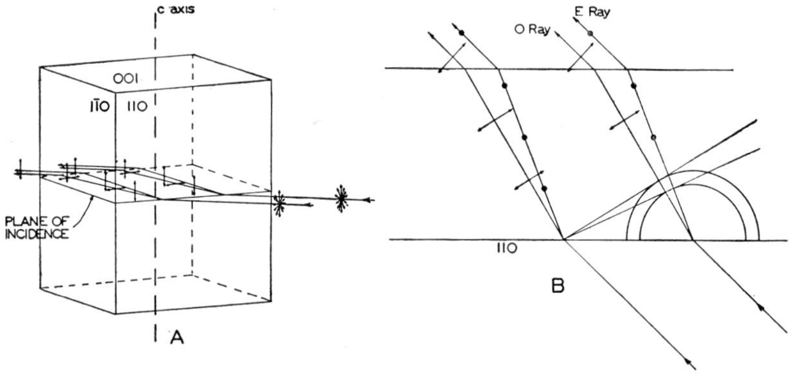

FIG. 11. Light falling with inclined incidence on a prism face of a positive tetragonal crystal. Plane of incidence normal to *c* axis (optic axis).

principal section. All monochromatic light passing through the crystal parallel to the optic axis travels with the same velocity.

Figure 9 illustrates nonpolarized light falling with perpendicular incidence on a prism face of an optically positive tetragonal crystal, the plane of incidence including the *c* axis. Upon entering the crystal,

the light does not deviate from its original path but is polarized into mutually perpendicular planes. The extraordinary component vibrates in a plane including the optic axis (the principal section) and the ordinary component in a plane perpendicular thereto. The extraordinary component in positive crystals travels more slowly than the ordinary component. The explanation is apparent upon examination of the Huygenian construction of Fig. 9*B*, which shows the ray velocity surfaces in a principal section.

Figure 10 illustrates another possibility. Light falling with inclined incidence on the basal plane of a positive hexagonal crystal is doubly refracted and polarized into mutually perpendicular planes. If the

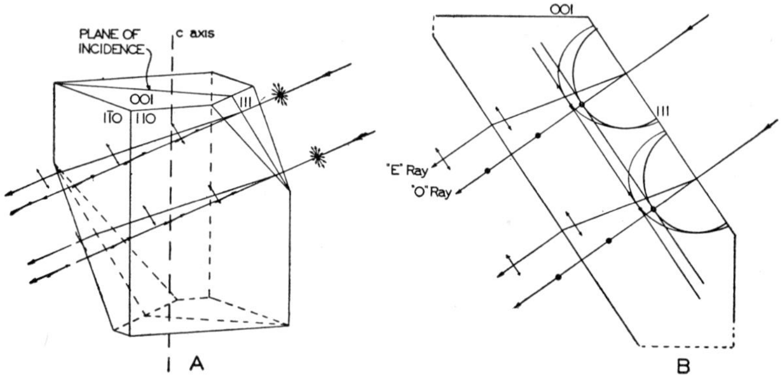

FIG. 12. Light perpendicularly incident on a pyramid face of a negative tetragonal crystal. Plane of incidence parallel to a principal section.

plane of incidence is a principal section, the relations shown in Fig. 10*B* are valid. The ordinary component is faster than the extraordinary component, as shown by the relative spacing of dots and arrows.

Light falling with inclined incidence on a prism face of a positive tetragonal crystal, the plane of incidence parallel to the basal pinacoid, undergoes double refraction and polarization as illustrated in Fig. 11. In a basal section the cross section of the ray velocity surfaces consists of concentric circles. The extraordinary component, dots in Fig. 11*B*, vibrates in the principal section and is slower than the ordinary component.

Figure 12 illustrates nonpolarized light perpendicularly incident on a pyramid face of a negative tetragonal crystal. The plane of incidence is assumed to be the principal section. The nature of double refraction and the relative velocities of the ordinary and extraordinary

components is determined by the Huygenian construction shown in Fig. 12*B*.

An example of light falling on a crystal in a random plane of incidence cannot be illustrated in a two-dimensional drawing. However, such light is doubly refracted into two mutually perpendicular planes. One component vibrates in the principal section, the other in a plane

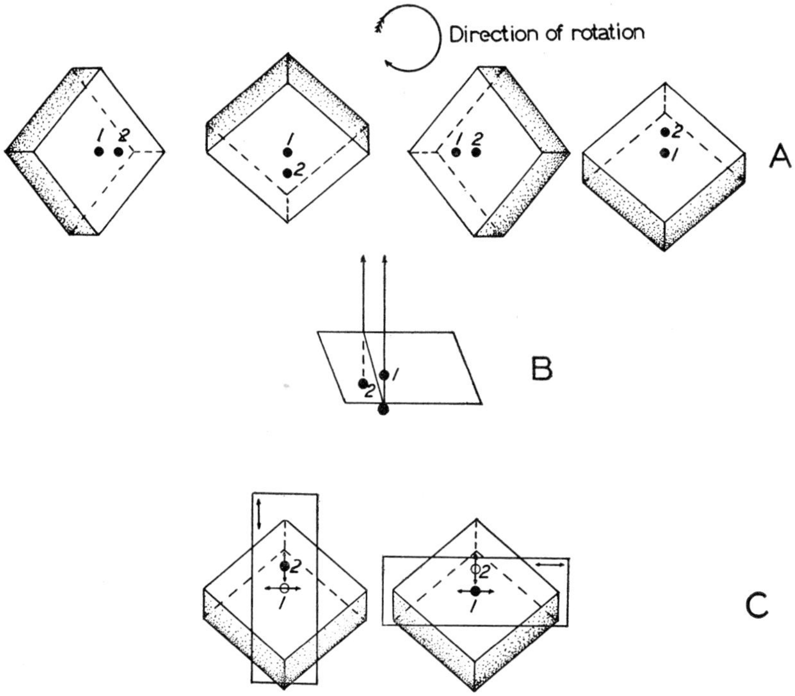

FIG. 13. The calcite experiment.

normal to the optic axis. The extraordinary component is slower or faster than the ordinary component, depending on whether the crystal is positive or negative.

In the Huygenian constructions shown in Figs. 8 to 12, the direction of a ray through a crystal is determined by drawing a line from a point of incidence of light on the surface of the crystal through the point of tangency of the wave front with the appropriate ray velocity surface. The angle of refraction determined in this manner when substituted in the formula

$$n = \frac{\sin i}{\sin r}$$

THE CALCITE EXPERIMENT   81

gives the ray index of refraction.   For light falling with inclined incidence on the face of a crystal the wave index of refraction may be
found by measuring the angle between a normal to the crystal face and
the wave normal in the crystal.   It should be noted that in Huygenian
constructions the extraordinary rays in general are *not* at right angles
to their corresponding wave fronts.

The Calcite Experiment.   Experimental proof of double refraction
is easily demonstrated with a clear calcite cleavage rhomb.   If a
calcite rhomb is placed over a dot on a piece of paper, two dots are
seen through the calcite, one at a shallower depth than the other.
When the rhomb is rotated, one dot revolves about the other, as is

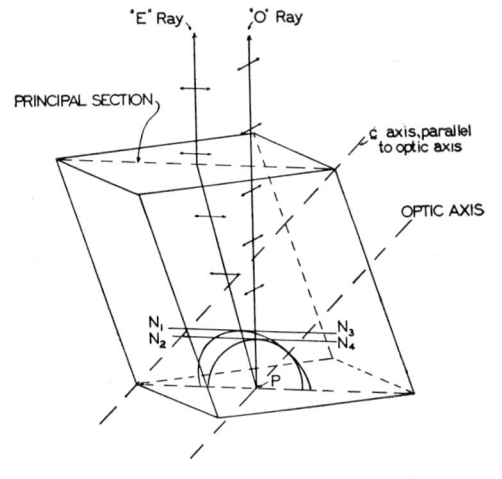

FIG. 14.   Passage of light through calcite.

shown in Fig. 13A.   Figure 13B shows in cross section the apparent
depth of each dot.   Light moving along the rays for the stationary
dot (1) passes through the calcite as if it were ordinary glass, but light
following the rays for the other dot (2) acts in an extraordinary
manner.

It can be demonstrated that the light traveling the two paths shown
in Fig. 13B is polarized in mutually perpendicular planes by superim-
posing a polaroid plate, a nicol prism, or a tourmaline plate over the
rhomb.   These devices permit light that is vibrating in one plane only
to pass through.   If the plane of polarization of the superimposed
plate parallels a principal section of the calcite, the revolving dot will
be seen, but the other dot is extinguished.   From this experiment it is
known that the refracted component is the extraordinary component.
If the plane of polarization of the superimposed plate is perpendicular

82   THE UNIAXIAL INDICATRIX

to the principal section of the calcite, only the stationary dot is seen. These experiments are indicated diagrammatically in Fig. 13C.

An explanation of the behavior of light in calcite is given in Fig. 14, which shows the ray velocity surfaces in a calcite rhomb in three dimensions. Light from a dot at P is doubly refracted into two components vibrating in mutually perpendicular planes. The ordinary component is constrained to vibrate in a plane normal to the principal section and passes without path deviation through the rhomb. The extraordinary component is bent and vibrates in the principal section. Although the extraordinary component travels with greater velocity than the ordinary component, its path in the calcite is longer than that of the ordinary component. The difference in the apparent depth of the images of the dots in the calcite depends on the fact that the light traveling along the extraordinary ray has a greater velocity than the light moving along the ordinary ray; that is, the calcite does not offer as much resistance to the passage of the extraordinary component as to the ordinary component. Stated another way, the index for the E component is less than the index for the O component.

Surfaces Related to the Indicatrix. Spatial variation of optical properties of crystals is usually discussed in terms of the indicatrix and the ray velocity surfaces. But other geometric and mathematical concepts may be derived from these surfaces when the nature of passage of light through crystals is considered from points of view other than those implied by the indicatrix and the ray velocity surfaces.

The *indicatrix* is a single-shelled surface—an ellipsoid of rotation. It is represented mathematically by the following formula:

$$\frac{x^2 + z^2}{n_O^2} + \frac{y^2}{n_E^2} = 1$$

where x, y, and z are the coordinates of conventional geometry.

The geometric relation of the *ray velocity surfaces* to the indicatrix has been discussed in a preceding section of this chapter. The ray velocity surface (ray surface) is two-shelled and consists of a sphere and an ellipsoid of rotation. The equation for these surfaces is

$$\frac{x^2 + z^2}{\left(\dfrac{1}{n_E}\right)^2} + \frac{y^2}{\left(\dfrac{1}{n_O}\right)^2} = 1$$

The *wave velocity surface* is derived from the ray velocity surface and consists of a rotation ovaloid and a sphere. Its relation to the ray velocity surface may not be apparent until certain fundamental

SURFACES RELATED TO THE INDICATRIX   83

considerations receive attention. In isotropic substances light travels outward from a point source of light with equal velocity in all directions and the ray velocity surface is a sphere. Because the light in each ray vibrates normal to the direction of propagation and the wave front is normal to each ray, a surface indicating wave velocity is a sphere also and coincides with the ray velocity surface. The same type of reasoning applies to the light moving along the ordinary rays in uniaxial crystals. That is, the ray velocity surfaces and the wave velocity surfaces are coincident spheres.

The wave velocity surface corresponding to the light moving along the extraordinary rays does not coincide with the ray velocity surface. This fact requires explanation. The vibration direction of light propagated along the extraordinary rays is in general not perpendicular to the direction of propagation but perpendicular to the wave

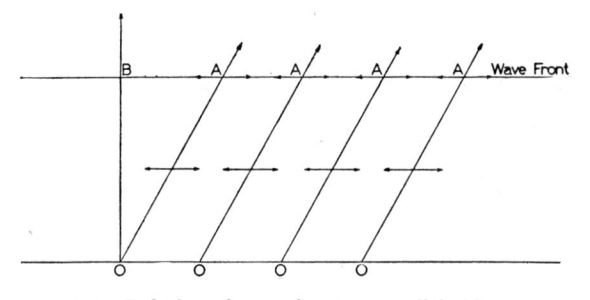

FIG. 15. Relation of wave front to parallel oblique rays.

normal. Consideration of this circumstance leads to the conclusion that the velocities of light in rays and waves differ. In Fig. 15 a wave advancing in the direction OB (the wave normal) may actually consist of light advancing along rays in the direction OA, if the vibration direction of the light in the rays is perpendicular to the wave normal but *not* perpendicular to the direction of propagation along the rays. Actually, in Fig. 15 the velocity of light moving along the rays exceeds the velocity of the wave in the direction of its normal. The velocity of light along the rays is measured by OA, but the velocity of the wave is measured by OB.

Figure 16 further illustrates these statements. Suppose that EO and $E'O'$ are parallel rays that mark the paths followed by nonpolarized light perpendicularly incident on the lower surface of a crystal plate, the plane of incidence lying in a principal section. If the plate is inclined to the optic axis, the light in each incident ray is split into two plane-polarized components. One, the ordinary component, passes through the plate without deviation. HH' is parallel to the wave

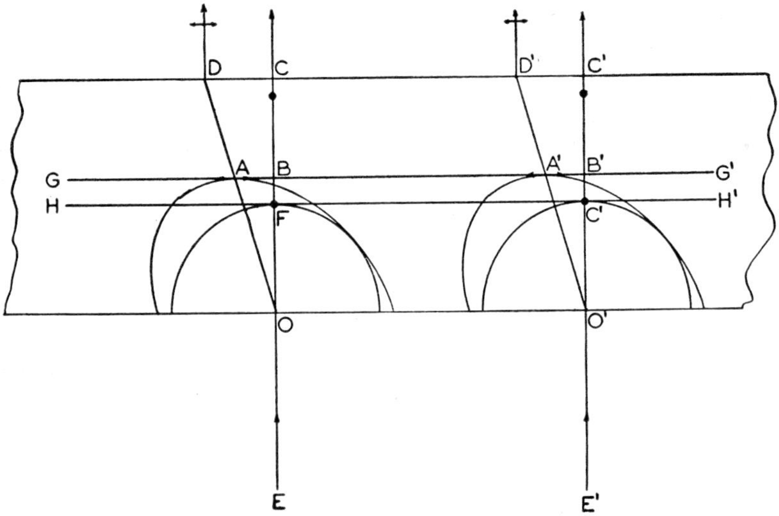

FIG. 16. Relation of wave fronts to direction of propagation of rays.

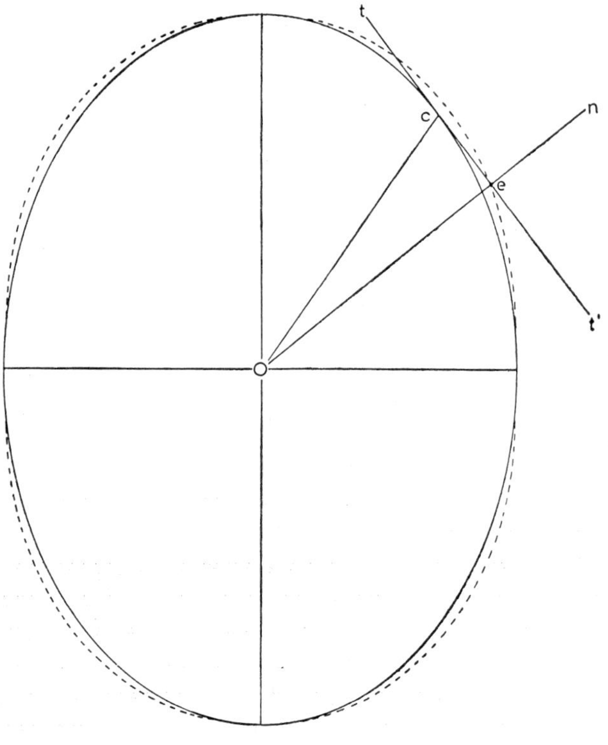

FIG. 17. Relation of wave velocity surface to ray velocity surface.

front for the ordinary components in the crystal plate. The other, the extraordinary component, is refracted and travels along rays OA and $O'A'$. The wave front for the extraordinary components is parallel to GG' which, in turn, is parallel to HH'. The velocity of the extraordinary component along its ray is measured by OA, but

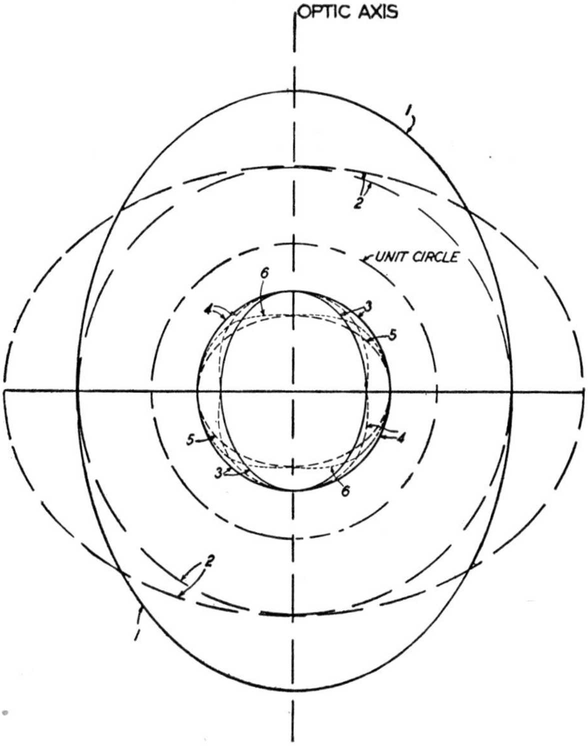

Fig. 18. Six related surfaces of uniaxial positive crystal.

1. Indicatrix.
2. Index surfaces.
3. Ray velocity surfaces.
4. Wave velocity surfaces.
5. Fresnel ellipsoid.
6. Ovaloid.

the velocity of the extraordinary wave in the direction of its normal is measured by OB. Thus the velocity of the light in the rays is greater than the velocity of the light in the wave.

In uniaxial crystals the ray velocity surface for light transmitted along the extraordinary rays is an ellipsoid of rotation. From the above considerations it can be demonstrated that for a point source the wave velocity surface for the extraordinary components is an ovaloid of rotation enveloping the ellipse. In Fig. 17, let the ellipse (solid

line) be a principal section of the ray velocity surface for light moving along the extraordinary rays of a uniaxial crystal. The extraordinary component of light traveling along ray Oc vibrates parallel to tt', the wave front. While light travels along the ray from O to c, the wave travels along its normal from O to e. Point e lies on an ovaloid (dashed line), which may be constructed by determining the wave velocities for light moving along all rays originating at O. In three dimensions the wave velocity surface constitutes an ovaloid of rotation whose axis is parallel to the optic axis.

The equation for the ovaloid of rotation is

$$\left(\frac{1}{n_E}\right)^2 (x^2 + z^2) + \left(\frac{1}{n_O}\right)^2 y^2 = (x^2 + y^2 + z^2)^2$$

The indicatrix is a single-shelled surface which shows the indices of refraction for waves in their *direction of vibration.* It is possible to draw a two-shelled surface which shows the variation of the indices of waves in their *direction of propagation.* The surface, which is called an *index surface,* consists of a sphere with radius n_O and of an ellipsoid of rotation with lateral and vertical axes equal respectively to n_E and n_O. The ellipsoid is defined by the equation

$$\frac{x^2 + z^2}{n_E^2} + \frac{y^2}{n_O^2} = 1$$

Derived from the ray velocity surface is the *Fresnel ellipsoid,* which shows the velocity of all light along rays in its *direction of vibration* (not direction of propagation as in the ray velocity surfaces). The Fresnel ellipsoid is a single-shelled ellipsoid of rotation with lateral and vertical axes respectively equal to $1/n_O$ and $1/n_E$. Its equation is

$$\frac{x^2 + z^2}{\left(\frac{1}{n_O}\right)^2} + \frac{y^2}{\left(\frac{1}{n_E}\right)^2} = 1$$

A sixth surface, single-shelled, is an ovaloid of rotation. It represents the *velocities* of waves in their *direction of vibration.* Its equation is

$$\left(\frac{1}{n_O}\right)^2 (x^2 + z^2) + \left(\frac{1}{n_E}\right)^2 y^2 = (x^2 + y^2 + z^2)^2$$

The various related surfaces for a positive crystal are shown in principal section in Fig. 18.

CHAPTER VIII

POLARIZATION OF LIGHT

Polarization by Reflection. Light is partly polarized by reflection; the amount of polarization is a function of the angle of incidence of the light, the index of refraction of the reflecting substance, and the quality of the reflecting surface. In Fig. 1, suppose that light is incident at

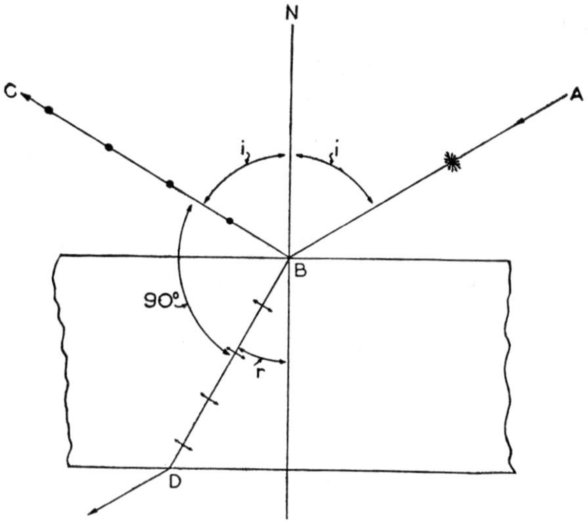

FIG. 1. Polarization of light by reflection and refraction. Brewster's law.

an angle of i on the reflecting surface of a plate of a substance. Part of the light is reflected through an angle equal to i, and part is refracted through an angle r. It has been proved experimentally that the reflected light is polarized most efficiently when the angle between the reflected and refracted light is 90 degrees (Brewster's law).

When this condition obtains,

$$n = \frac{\sin i}{\sin r} = \frac{\sin i}{\sin (90 - i)} = \tan i$$

where n is the index of refraction of the reflecting substance.

In Fig. 1 the polarized reflected light in ray BC vibrates parallel to the reflecting surface. The refracted light in ray BD is polarized

in a plane including the ray for the incident light and the normal to the surface at the point of incidence. It should be emphasized that the reflected and refracted light is not completely plane-polarized. However, repeated reflection results in light with a high degree of plane polarization.

Polarization by Absorption. Certain anisotropic substances, such as tourmaline, absorb transmitted light much more strongly in one direction than in another. Nonpolarized light entering tourmaline is split into two components, one vibrating in the principal section (the E component) and the other in a plane perpendicular to the principal section (the O component). The O component is strongly absorbed, and

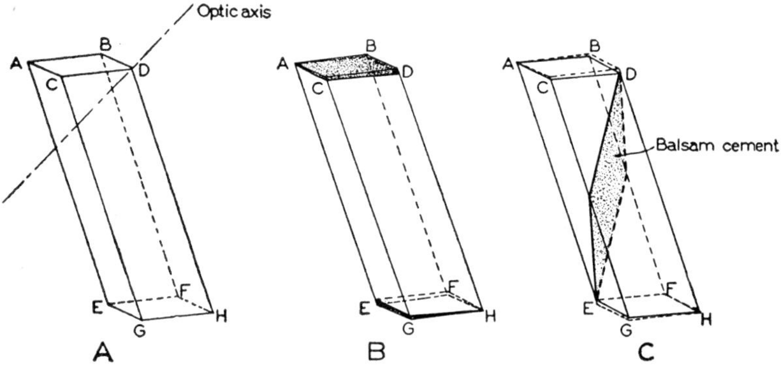

FIG. 2. Steps in construction of simple nicol prism.

A. Calcite cleavage rhomb.
B. Cleavage rhomb showing portion ground away (stippled).
C. Cut and recemented rhomb, completed prism.

only a small fraction passes through. The E component, however, is not impeded appreciably and passes through the crystal easily. The result is that nearly all the transmitted light is polarized in a plane paralleling the c crystallographic axis of the tourmaline.

Polaroid, a manufactured substance widely used in sun glasses and various types of optical equipment, owes its ability to produce plane-polarized light to differential absorption of transmitted light.

Polarization by Double Refraction. Light passing through anisotropic substances in any direction except that of an optic axis is plane-polarized. The emergent light consists of two components vibrating in mutually perpendicular planes.

Strong double refraction of light by calcite has led to its use in the construction of polarizing prisms. The construction of the *nicol* prism, the first type to be used in petrographic microscopes, is indicated in

Fig. 2. A flawless piece of clear calcite, "Iceland spar," is split to produce an elongated cleavage rhomb about three times as long as it is broad (Fig. 2*A*). The end faces, which naturally meet the edges *AE* and *DH* at angles of 70° 53′, are ground so that the angles become 68 degrees (Fig. 2*B*). The calcite is then sawed diagonally, as in Fig. 2*C*, at right angles to the ground and polished end faces. The halves are cemented together with balsam and the sides of the prism are covered with an opaque coating.

The passage of light through a nicol prism, or "nicol," is diagrammatically indicated in Fig. 3, a cross section. Light moving along ray *MN*, upon striking the lower surface of the nicol, is doubly refracted into two components which vibrate in mutually perpendicular planes. Inasmuch as the index of refraction for the ordinary component in calcite is 1.658 and the index of the balsam film is about 1.54, the ordinary component *NOP* is totally reflected at the contact of the calcite with the balsam film and passes into the walls of the prism where is it absorbed. The extraordinary index n_E of calcite is 1.486, and light traveling in intermediate positions between the component with the minimum index and the optic axis has an index whose value is somewhere between 1.658 and 1.486. It so happens that the extraordinary component traveling parallel or nearly parallel to the long direction of the prism has an index in the neighborhood of 1.54, the index of balsam.

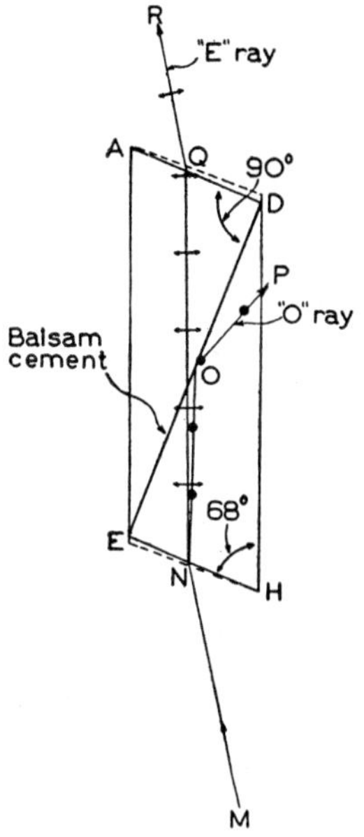

FIG. 3. Cross section of nicol prism, showing paths of transmitted light.

Therefore, the extraordinary component travels directly through the balsam film and emerges from the upper surface of the prism as plane-polarized light.

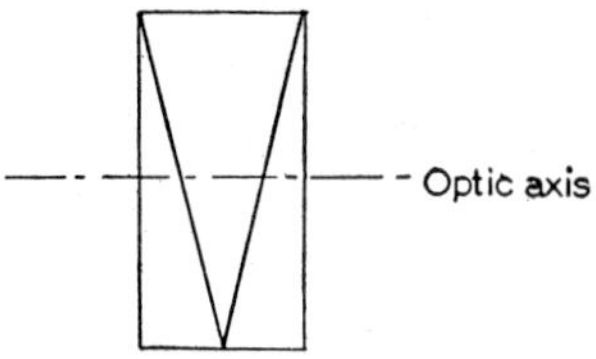

FIG. 4. Ahrens' prism.

Since the nicol prism was invented, many other types of polarizing calcite prisms have been constructed. One of the most popular prisms in use today is *Ahrens'* prism, a cross section of which is shown in Fig. 4. It consists of three segments properly cut and cemented together by balsam. In this prism the ordinary components are reflected to both sides, and the extraordinary component emerges to produce plane-polarized light.

The effective angle between extreme rays transmitting emergent light is greater than in the nicol prism.

Polarization by Scattering. This phenomenon is of slight importance in optical crystallography and is given only brief notice here. Light waves cause small particles in suspension in a liquid or gas, or even the molecules of the medium themselves, to serve as sources of secondary "scattered" radiation which spreads out from the particles in all directions. The "scattered" light is partly plane-polarized, in that it has a stronger component of transverse vibration in one direction than in other directions. Almost all of the light reaching the earth from the sky is the result of scattering by the upper atmosphere and is partly plane-polarized, a fact that can be

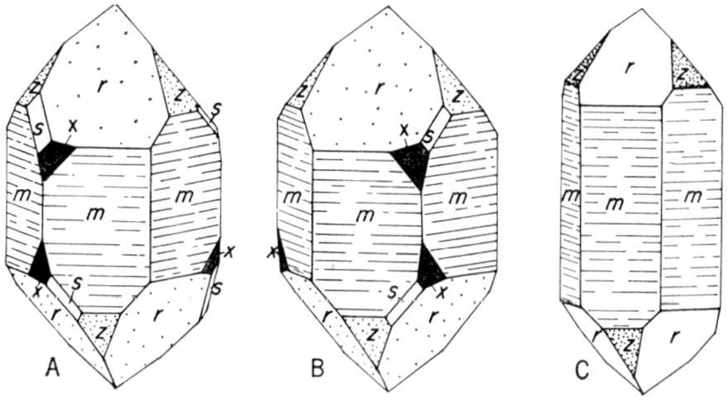

FIG. 5. Quartz crystals.

A. Left-handed quartz.
B. Right-handed quartz.
C. Typical elongate crystal.

tested by observing sky light directly above the observer through a polarizing prism. Inasmuch as the short wave lengths of white light are scattered more than the long ones, the sky appears blue.

Rotation of the Plane of Polarization. Crystals in at least twelve of the thirty-two crystal classes possess the ability to rotate the plane of polarization of light transmitted along an optic axis in anisotropic crystals, or in any direction in isotropic crystals, and are said to be *optically active.* The ability to rotate the plane of polarization is measured by the *rotatory power,* and the phenomenon is described as *rotary polarization* or *rotatory polarization.* Crystals that display this effect are found in each of the six crystal systems; they have certain symmetry characteristics in common, notably the absence of a center of symmetry and few or no planes of symmetry. Enantiomorphism is evident in that left- and right-handed forms are common.

Quartz, in the trigonal trapezohedral class, illustrates rotary polari-

ROTATION OF THE PLANE OF POLARIZATION   91

zation.   Quartz crystals are shown in Fig. 5 and may be either left-handed or right-handed.   If a section is cut at right angles to the c axis and plane-polarized light is passed through the crystal parallel to the c axis, the apparent result is something like that in Fig. 6, which shows a basal plate of left-handed quartz.   It is not believed

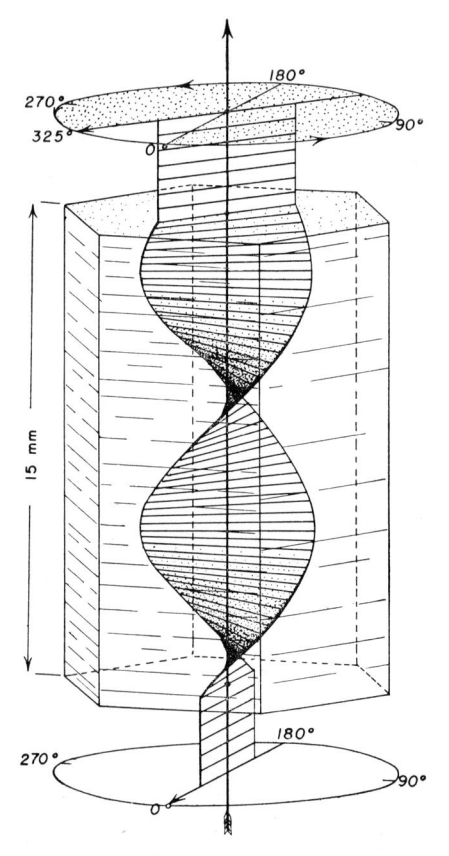

FIG. 6.   Rotation of plane of polarization of polarized monochromatic light by basal plate of left-handed quartz.

that the plane of polarization is actually rotated inside the crystal plate as shown in Fig. 6, but it is known that the amount of apparent rotation of the plane of polarization of the light leaving the plate with reference to the plane of polarization of incident light is a direct function of the thickness of the plate.   For the D wave length of the visible spectrum a basal plate of quartz 1-mm thick rotates the plane of polarization 21° 40′ at 20° C.

Left-handed quartz rotates the plane of polarization to the left as

92   POLARIZATION OF LIGHT

one looks along the c axis against the direction of the advancing light; it is said to be *levo-rotatory*. Right-handed quartz is *dextro-rotatory*. A common explanation of rotary polarization assumes that the crystal separates the incident plane-polarized light into two components, circularly polarized in opposite directions, and that these components travel through the crystal with different velocities. Upon emergence, one component will have rotated through a larger angle than the other. The resultant of the combination of the two components gives plane-polarized light the vibration plane of which is rotated somewhat with respect to the plane of the incident light.

Rotary polarization is generally omitted from the discussion of the theory of optical crystallography for the reason that most optical measurements are made on thin sections or fragments or crystals in which the rotation of the plane of polarization is so slight that it can be regarded as insignificant. Also most crystalline substances belong to crystal classes which do not exhibit rotary polarization.

CHAPTER IX

UNIAXIAL CRYSTALS IN PLANE–POLARIZED LIGHT

Introduction. The presence of polarizing plates or prisms in a petrographic or chemical microscope distinguishes it from an ordinary biological microscope and makes it an ideal instrument for the study of the manner of passage of light through crystal plates or fragments. In a petrographic microscope the light passing through the lower polarizing prism and entering a crystal is vibrating in one direction only; moreover, if the upper prism is inserted, the light leaving the crystal, before it reaches the eye, is again constrained to vibrate in a single direction. Ordinarily the planes of polarization of the upper and lower prisms are in a fixed position 90 degrees apart.

Interference Colors. Anisotropic nonopaque crystal fragments or plates between crossed polarizing prisms are characterized by interference colors, the nature and intensity of which depend on (1) the orientation of the fragment; (2) the thickness; (3) the birefringence (the difference between the maximum and minimum refractive indices); and (4) the nature of the light absorbed and transmitted by the fragment. The last-named factor is of no importance in uncolored substances.

If an uncolored anisotropic crystal fragment of uniform thickness is rotated on the microscope stage between crossed polarizing prisms, it shows a particular color which changes in intensity from a maximum to a minimum four times during a 360-degree rotation. The positions occupied by the fragment at minimum illumination (usually complete darkness) are described as the *extinction positions*.

Figure 1 shows diagrammatically the interaction of the petrographic microscope and a uniaxial crystal plate. Monochromatic nonpolarized light entering the lower polarizing prism (polarizer) emerges vibrating in one plane only. This light strikes the lower surface of the crystal plate, where in general it is doubly refracted and resolved into two components vibrating in mutually perpendicular planes. Upon emergence, the extraordinary component is again refracted so that its path becomes parallel to that of the ordinary component. Both components, upon passing through the upper polarizing prism (analyzer), are again doubly refracted, but only the light vibrating in the plane of polarization of the upper polarizing prism passes through.

If monochromatic light passes through the crystal plate, constructive or destructive interference takes place, and light or darkness results. Whether light or darkness results depends on the path difference between the ordinary and extraordinary components produced by the crystal plate and on the orientation of planes of polarization of the crystal with respect to the planes of polarization of the polarizing prisms.

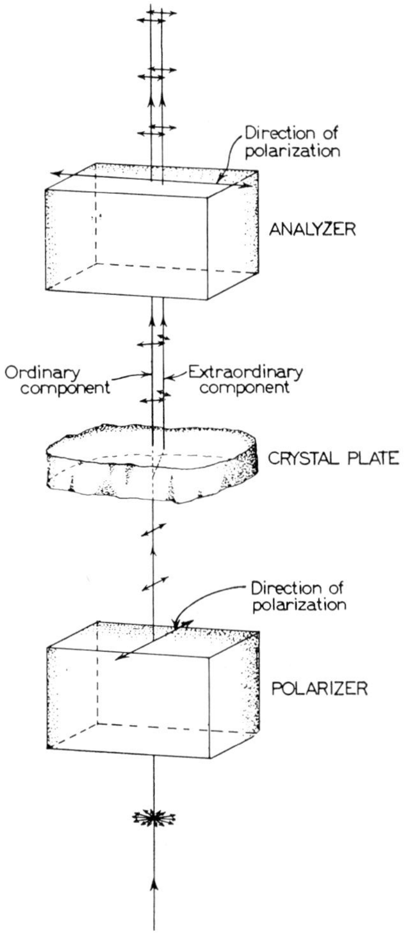

FIG. 1. Passage of light through crystal plate in polarizing microscope.

Graphic analyses of the interaction of a petrographic microscope and a crystal plate are shown in Figs. 2 to 5.

In Fig. 2 plane-polarized monochromatic light vibrating in a plane parallel to PP' enters a crystal plate rotated 45 degrees (Fig. 2*A*) or 15 degrees (Fig. 2*B*) from an extinction position. The light entering the crystal is resolved into two components vibrating in mutually perpendicular planes, the traces of which are XX' and YY'. Inasmuch as the two components travel with different velocities through the crystal plate, the emergent waves in general will be out of phase and will interfere so as to give a plane-polarized, circularly-polarized, or elliptically-polarized resultant. In Fig. 2 no effort has been made to show the resultant of the combination of the waves emerging from the crystal plate. Instead, the waves are left uncombined so as to show to what extent the crystal plate has produced a path difference (phasal difference).

Suppose that, as in Fig. 2, the crystal plate has produced a path difference of $n\lambda$, where n is a whole number such as 1, 2, 3, etc., and λ is the wave length. The wave resulting from the combination of the light leaving the crystal plate enters the upper (analyzing) prism and

is resolved into two components, only one of which can pass through. The light emerging from the upper prism vibrates in a plane, the trace of which is AA', at right angles to the vibration plane of the lower

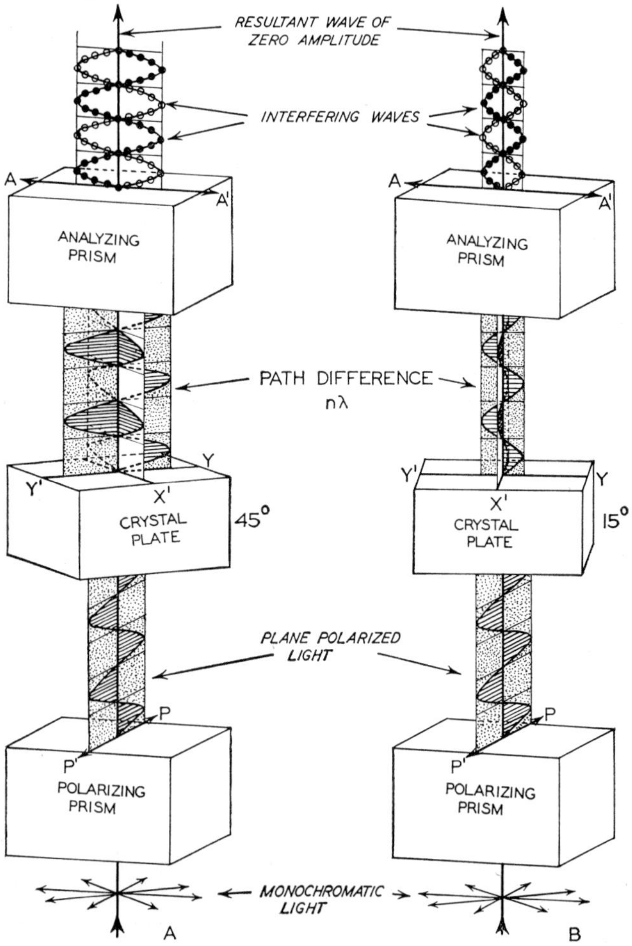

FIG. 2. Passage of monochromatic light through crystal plate between crossed polarizing prisms. Path difference $n\lambda$.

A. 45-degree position.
B. 15-degree position.

prism. If the path difference produced by the crystal plate is $n\lambda$, the waves emerging from the upper prism are $(n/2)\lambda$ out of phase, where n is an odd whole number; and because the waves are similar and oppose each other, the result is a wave of zero amplitude. The

waves leaving the upper prism are $(n/2)\lambda$ out of phase because the planes of polarization of the upper and lower prisms are mutually perpendicular.

Figure 3*A*, a vector diagram, is helpful in explaining this phenomenon. PP' and AA' are the traces of the vibration planes in the lower and upper polarizing prisms assuming that the observer is looking down into the microscope. OB is the amplitude of the wave leaving the lower prism, and XX' and YY' are the traces of the vibration planes on the crystal plate. In the 45-degree position the amplitudes of the two components passing through the crystal plate are equal and, if the plate produces a path difference of $n\lambda$, are obtained by dropping

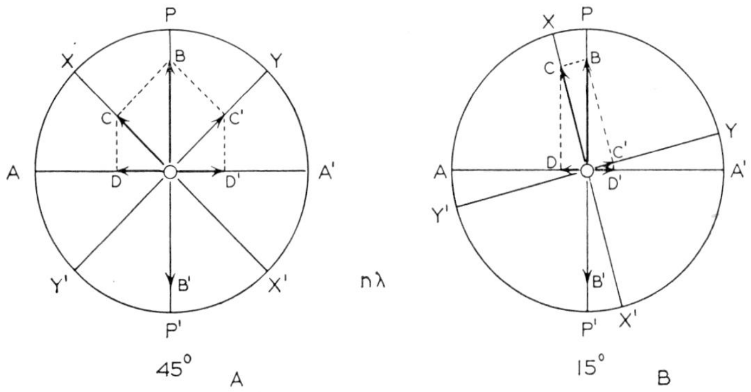

FIG. 3. Action of petrographic microscope on crystal plate in a vector diagram. Plan view. Path difference $n\lambda$.

A. 45-degree position.
B. 15-degree position.

normals to XX' and YY' from B. OC and OC' are found in this manner. The waves leaving the crystal plate combine and are resolved again in the upper prism. The amplitudes of the components passing through the upper prism are obtained by dropping normals to AA' from C and C'. Observe that OD and OD', thus derived, are equal and opposite and yield a resultant wave of zero amplitude.

In the 15-degree position (Figs. 2*B* and 3*B*) the components emerging from the crystal plate have unequal amplitudes, but the light leaving the upper prism still gives a resultant wave of zero amplitude. Note that the opposing emergent waves have smaller amplitudes than the waves emerging from the upper prism when the crystal plate is in the 45-degree position.

Figures 4 and 5 portray the passage of monochromatic light through a crystal plate in such a manner that the emergent waves have a path

difference (phasal difference) of $(n/2)\lambda$, where $n = 1, 3, 5, 7$, etc., and λ is the wave length. Again the wave resulting from combination of the waves emerging from the crystal plate is not shown; however,

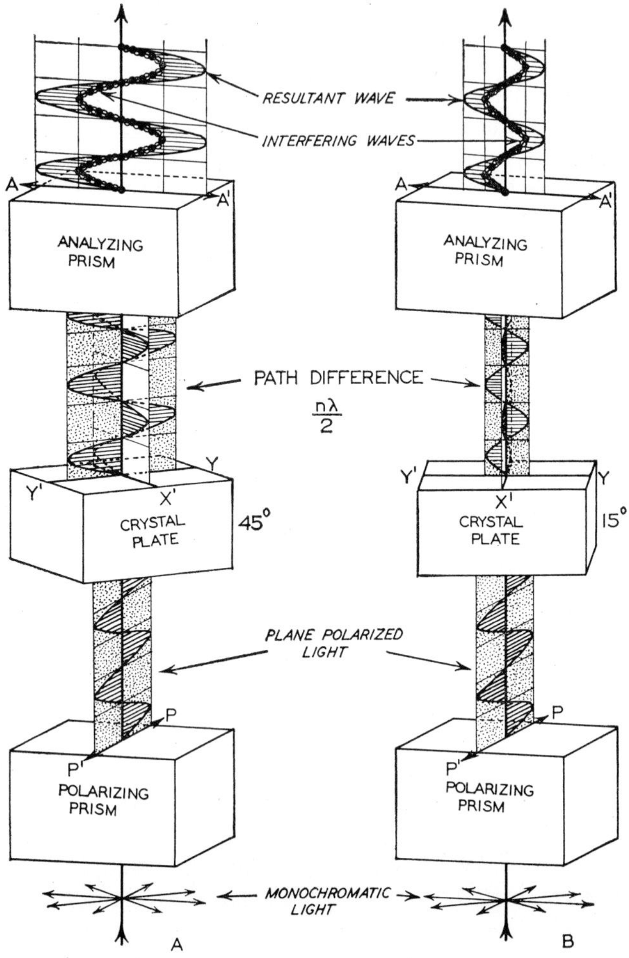

FIG. 4. Passage of monochromatic light through crystal plate between crossed polarizing prisms. Path difference $(n/2)\lambda$.

A. 45-degree position.
B. 15-degree position.

this omission does not detract from the validity of the argument, because the light is again resolved when it enters the upper polarizing prism.

The light waves leaving the upper prism are in phase and super-

imposed so that a resultant wave of twice the amplitude of either of the interfering waves is obtained. The light from this wave is four times as intense as the light from either of the interfering waves for the reason that intensity varies as the square of the amplitude of a wave.

The resultant wave leaving the upper polarizing prism has a maximum amplitude when the vibration planes of the crystal plate are in the 45-degree position, as in Fig. 4*A*. This is explained by reference to Figs. 5*A* and *B*. In Fig. 5*A*, PP' and AA' are the traces of the vibration planes of the lower and upper polarizing prisms. XX' and YY' are the traces of the vibration planes of the crystal plate. OB is the

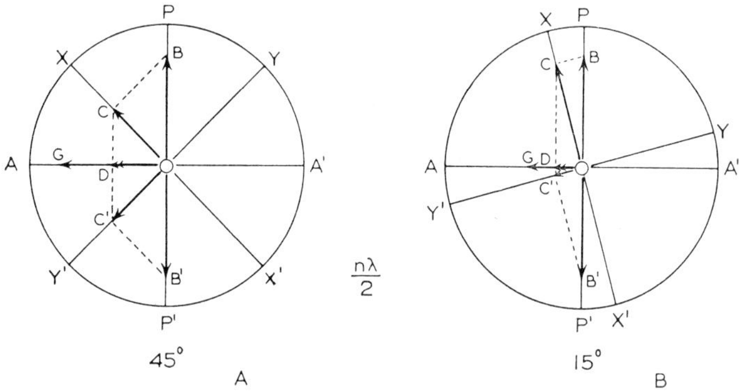

FIG. 5. Action of petrographic microscope on crystal plate in a vector diagram. Plan view. Path difference $(n/2)\lambda$.

A. 45-degree position.
B. 15-degree position.

amplitude of the wave entering the crystal plate. If the crystal plate produces a path difference of $(n/2)\lambda$ the amplitudes of the components leaving the crystal plate can be obtained by dropping normals from B and B' to XX' and YY'. OC and OC' represent the waves which are resolved again when passing through the upper prism. Two waves emerging from the upper prism act in the same direction and have amplitudes equal to OD. The resultant wave has an amplitude OG equal to twice OD.

A similar construction for the 15-degree position (Fig. 5*B*) indicates that the amplitude OD of light in the rays emerging from the upper prism is a function of the angle between the vibration planes of the prisms and the crystal plate. The resultant wave has maximum amplitude in the 45-degree position and reduces to a wave of zero amplitude in the parallel position. Thus, it is seen that a crystal

plate will extinguish four times during a 360-degree rotation as the vibration planes of the crystal and the polarizing prisms become coincident.

Any fractional path difference produced by the crystal plate causes illumination, but the intensity of the light decreases as the fractional difference approaches unity or a multiple of unity. Moreover, no matter what the intensity of the transmitted light is, the illumination decreases to zero as the planes of polarization of the crystal approach coincidence with the planes of polarization of the upper and lower polarizing prisms.

If white light is passed through a crystal plate of constant thickness, a path difference of $(n/2)\lambda$ for some wave length of the spectrum results, and the color corresponding to that wave length is seen. One wave length (that of the complementary color) may have a path difference of $n\lambda$ and is removed from the white light by the microscope.

The crystallographic orientation of a fragment with respect to the axis of the microscope has an important bearing on its interference color. If a crystal fragment is oriented so that its optic axis is parallel to the axis of the microscope, no interference color is produced because all light traveling through the crystal has the same velocity, and no path difference results. A crystal fragment of the same thickness, lying so that its optic axis is normal to the axis of the microscope, produces a maximum interference color, because light traveling at right angles to the optic axis is resolved into components with maximum path difference.

A fragment with an intermediate orientation shows an interference color somewhere between the maximum and minimum colors.

Two crystal plates differing only in thickness have different interference colors because they produce unlike path differences. The thicker plate shows a color higher in the spectrum than the thinner plate. The principle operating here is illustrated by the quartz wedge —an elongate wedge cut from clear quartz and usually oriented so that its longer and shorter dimensions are parallel to the traces of the planes of vibration of the faster and slower components in the quartz, respectively.

If a monochromatic yellow-light source is used and the wedge is inserted into the microscope, thin edge first, light bands are seen where the path difference produced by the wedge is $(n/2)\lambda$, where $n = 1, 3, 5$, etc. The light is cut out where the path difference is $n\lambda (n = 1, 2, 3$, etc.). The wave length of average yellow light is about 580 mμ (1 mμ equals one millionth of a millimeter). Where the path

difference in millimicrons produced by the quartz wedge is 580/2, 3·580/2, 5·580/2, etc., yellow bands appear; where the path difference produced by the wedge is 580, 2·580, 3·580, etc., dark bands appear.

In Fig. 6 suppose that A is a ray marking the path of yellow light striking the quartz wedge at the thin edge. No path difference is produced; accordingly, between crossed polarizing prisms no light is seen under the microscope. Light in ray B, however, is doubly refracted, and a path difference of 290 mμ is produced. This path difference corresponds to one-half wave length for yellow light, and light is transmitted by the microscope. Light in ray C travels a path such that the quartz wedge causes a path difference of 580 mμ, or one wave length.

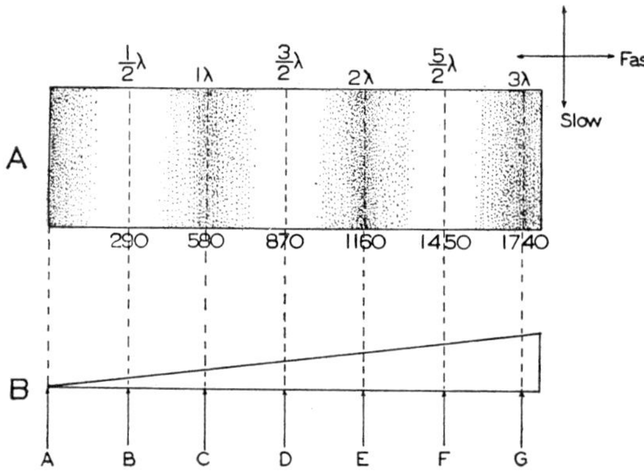

FIG. 6. Quartz wedge in monochromatic light between crossed polarizing prisms. Stippled areas dark.

A. Plan. *B*. Cross section.

Light traveling along this path gives a final resultant wave of zero amplitude. The same condition holds for light traveling along rays E and G. The total result is a series of alternating black and yellow bands, an effect which depends on the varying thickness of the quartz section.

The quartz wedge resolves white light into its spectrum. Before this is understood, it is necessary to consider again the nature of white light.

White light has no definite wave length but consists of a combination of light waves with wave lengths ranging from about 390 mμ to 770 mμ. The shortest wave lengths are those of violet, which has wave lengths ranging from 390 mμ to 430 mμ. Red wave lengths range from 650 mμ to 770 mμ and are the longest. The wave lengths for the

component colors of white light are indicated in the accompanying table.

WAVE LENGTHS OF VISIBLE LIGHT
Expressed in Millimicrons

	Range	*Average*
Violet	390–430	410
Indigo	430–460	445
Blue	460–500	480
Green	500–570	535
Yellow	570–590	580
Orange	590–650	620
Red	650–770	710

For the purpose of this discussion it is sufficient to use an average wave length for each color. Figure 7 shows the distribution of color bands in the quartz wedge for each of the colors listed above. The

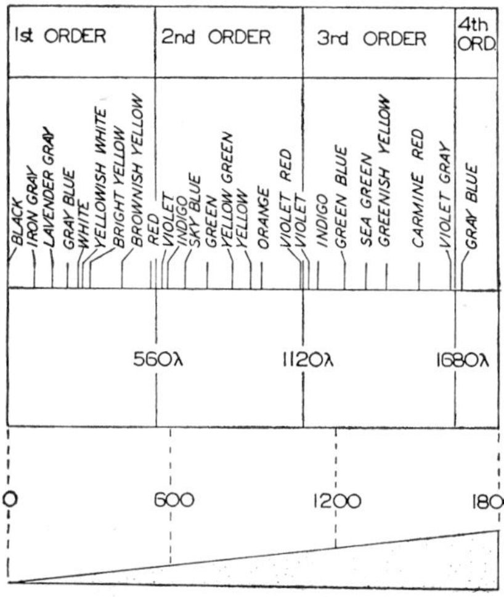

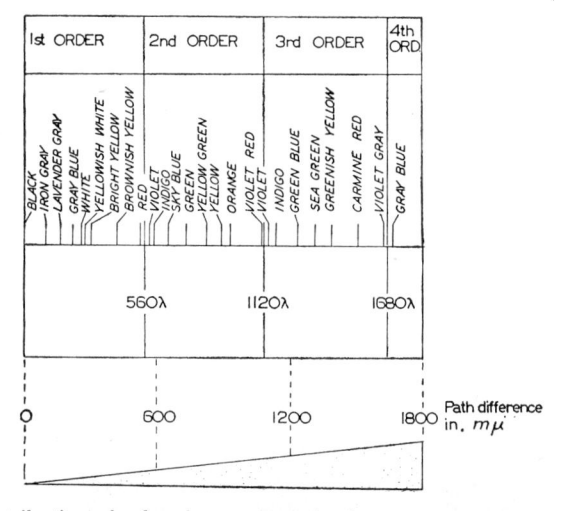

FIG. 7. Distribution of colors in quartz wedge between crossed polarizing prisms.

effect of the quartz wedge on white light is the summation of the effects for the individual colors, as diagrammatically shown in Fig. 8. As the quartz wedge is inserted into the petrographic microscope, thin edge first, the colors change from black through gray, white, and yellow to red, which marks the highest color of the *first-order spectrum.* The colors in the lower part of the first-order spectrum are, for the most part, a mixture of the various colors. This is understood by reference to Fig. 8. The *second-order spectrum* is more sharply separated into its component colors. Violet is followed by indigo,

blue, green, yellow, orange, and red. The colors of the *third-order spectrum* are less pronounced but nevertheless quite apparent.

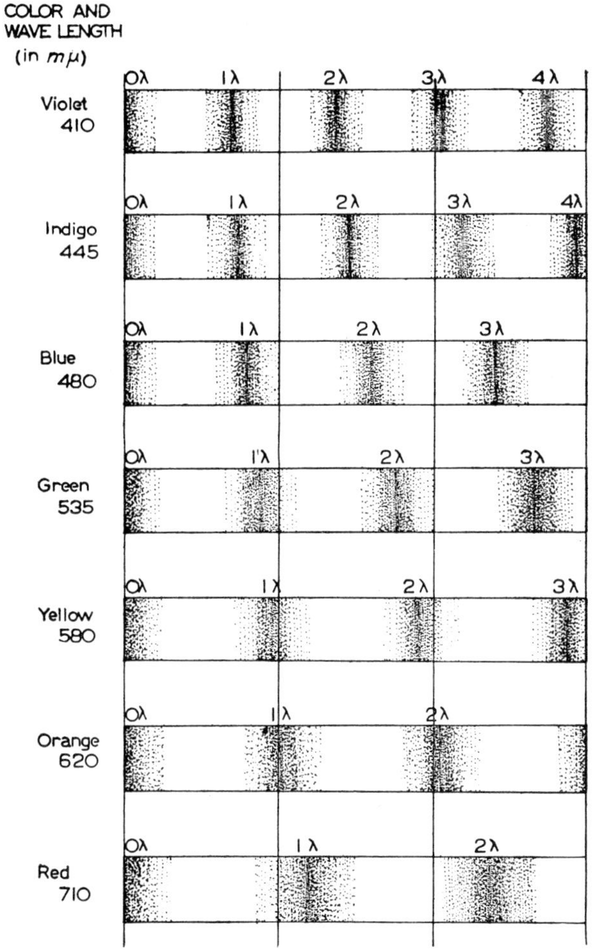

FIG. 8. Distribution of monochromatic color bands in quartz wedge between crossed polarizing prisms. Stippled areas dark; clear areas light.

The fourth, fifth, and higher orders show a peculiar white color resulting from the mixing of the component colors of white light. This white is called *white of a higher order.*

Birefringence. The *birefringence* of a crystal is the numerical difference between the maximum and minimum indices of refraction. The interference color of a crystal fragment or plate between crossed polarizing prisms depends in part on the birefringence, as is shown

by the fact that two crystal plates having identical thicknesses and orientations but different birefringences give different interference colors. The plate with higher birefringence yields the higher color. The velocities of the ordinary and extraordinary components vary as the reciprocals of their indices. Consequently, as the birefringence of crystal plates of constant thickness and orientation increases, the path differences produced by the plates increase.

The interrelationships of thickness, birefringence, and interference colors may be seen in Fig. 9. This diagram shows interference colors and the path differences in millimicrons necessary to produce them (abscissae), thickness in thousandths of a millimeter (ordinates), and birefringence (diagonal lines).

Quartz, which has a birefringence of 0.009, serves to illustrate the use of the diagram. Suppose that a plate of quartz shows a maximum interference color of yellowish white, and it is desired to determine the thickness of the plate. The diagonal line for birefringence 0.009 is followed with the eye until it intersects the vertical line for a path difference corresponding to yellowish white. The point of intersection projected to the left on a horizontal line gives the thickness of the fragment (in this case 0.030 mm).

Suppose that it is desired to determine the birefringence of a crystal fragment of known thickness and displaying a maximum interference color. Assume that the thickness is 0.040 mm, and the interference color is blue of the second order. The horizontal line for thickness 0.040 mm is followed to the right until it intersects a vertical line corresponding to a path difference necessary to produce second-order blue. The diagonal line passing through the point of intersection gives the birefringence, which turns out to be 0.016.

Determination of the Order of an Interference Color. It may be necessary at times to determine the order of an interference color. This is easily accomplished if a crystal fragment has a wedge edge which shows the colors of the spectrum in sequence from the edge toward the center of the fragment. Red bands are conspicuous and mark the upper limits of the colors in each order. Suppose that the color of a crystal fragment is predominantly blue and two red bands are seen in the wedge edge of the fragment. The lower red band marks the upper limit of the first-order colors, and the higher red band limits the colors of the second-order spectrum. Accordingly, the blue must be a third-order color.

If a crystal plate is uniformly thick, it becomes necessary to use a quartz wedge to determine the order of the interference color. The crystal plate is turned to the extinction position, where the traces of

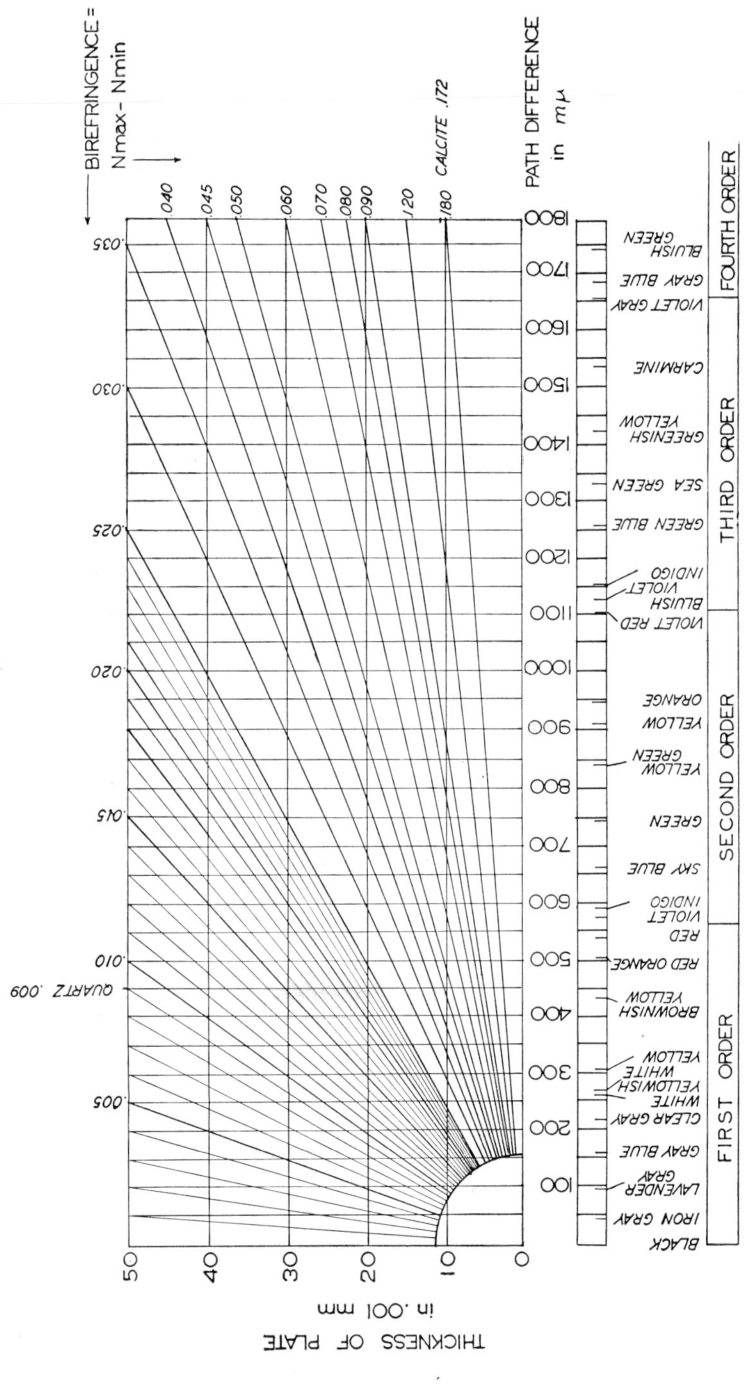

FIG. 9. Diagram showing relations of interference colors to thickness and birefringence in crystal plates.

the mutually perpendicular vibration planes in the crystal are parallel to the planes of polarization of the upper and lower polarizing prisms. The crystal fragment then is rotated 45 degrees so that the trace of the plane of vibration of either the slower or faster component is parallel to the long direction of the quartz wedge as it is inserted in the microscope. Suppose that the quartz wedge is constructed so that its fast component (ordinary component) vibrates in a plane parallel to the long direction of the wedge. If the direction of the trace of the plane of vibration of the fast component in the crystal is parallel to the direction of the trace of the plane of vibration of the fast component in the wedge, the interference color in the crystal fragment goes up as the wedge is pushed in, thin edge first. The wedge in effect adds to the thickness of the fragment and increases the path difference.

If the trace of the vibration plane of the slow component in the crystal fragment is parallel to the trace of the vibration plane of the fast component in the wedge, the colors in the fragment go down in the spectrum as the wedge is inserted. With slow and fast components parallel, darkness results when the path difference produced by the crystal fragment exactly equals the path difference produced by the wedge. When this occurs, the wedge is said to *compensate* the interference color of the fragment.

Suppose that a crystal fragment has a yellow interference color. The fragment is turned 45 degrees from extinction, and the quartz wedge is inserted thin edge first. If the interference color changes to red, then violet, the traces of the vibration planes of the fast components in the fragment and the wedge are parallel, and the crystal fragment should be rotated 90 degrees. Now as the wedge is inserted, the interference colors go down the scale. If the colors appear in the order yellow, green, blue, violet, red, orange, yellow, white, gray, and black—the yellow is a second-order color. If the crystal fragment is removed and the color is observed in the wedge alone, the colors listed above will appear in the same order as the wedge is pulled out.

The mechanism of compensation is indicated diagrammatically in Figs. 10 and 11. In Fig. 10, monochromatic light, incident on a crystal plate rotated 45 degrees from the extinction position, is resolved into two components vibrating in mutually perpendicular planes. For the sake of simplicity, it is assumed that the crystal plate produces a path difference of $\frac{1}{2}\lambda$ so that the light emergent from the plate gives a resultant plane-polarized wave of the same amplitude as the incident wave but rotated 90 degrees. In general, the resultant of the light

emerging from the crystal plate is circularly or elliptically polarized, but this light is resolved again upon entering the quartz wedge.

The light entering the quartz wedge is resolved into two components which vibrate in the mutually perpendicular vibration planes of the wedge. If the trace of the vibration plane of the fast component in the quartz wedge is parallel to the trace of the vibration plane of the slow component in the crystal plate, the effect is a

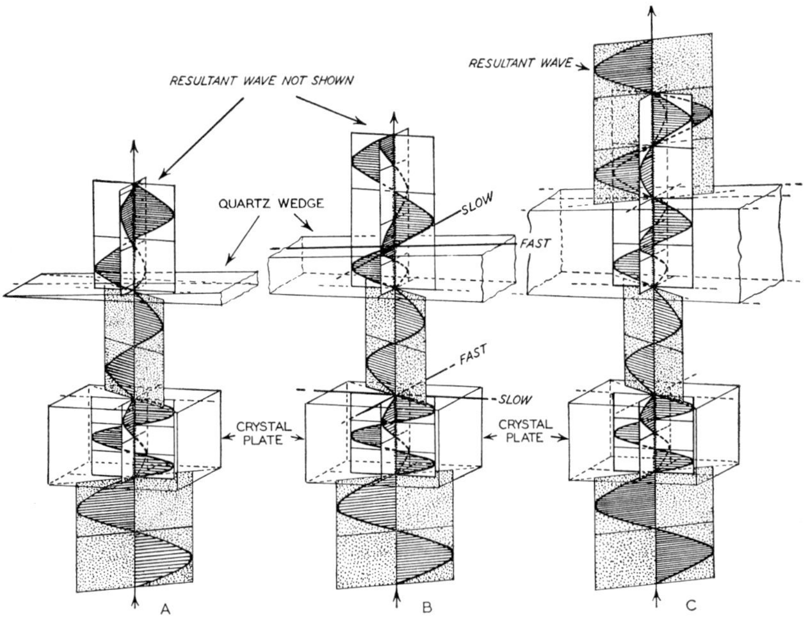

Fig. 10. Diagram showing effect of insertion of quartz wedge over a crystal plate which has produced a path difference of one-half wave length for the two components of monochromatic light. Fast direction of quartz wedge is parallel to the slow direction of the crystal plate. Compensation has been effected in position *C*.

reduction of the path difference produced by the crystal plate, and the waves leaving the quartz wedge have a smaller path difference than the waves leaving the crystal plate. In Fig. 10*A* the quartz wedge has reduced the path difference only slightly, in Fig. 10*B* more so, and in Fig. 10*C* the thickness of the wedge is such that it restores the light to the condition that characterized it when it was first incident on the crystal plate. At this thickness the quartz wedge has *compensated* the color produced by the crystal plate, and between crossed polarizing prisms the result is darkness.

The resultant wave for the light emergent from the quartz wedge

in Figs. 10*A* and *B* is not shown; both the emergent waves and their resultant are indicated in Fig. 10*C*.

Figure 11 is a two-dimensional drawing showing compensation by a quartz wedge. Monochromatic plane-polarized light incident on a

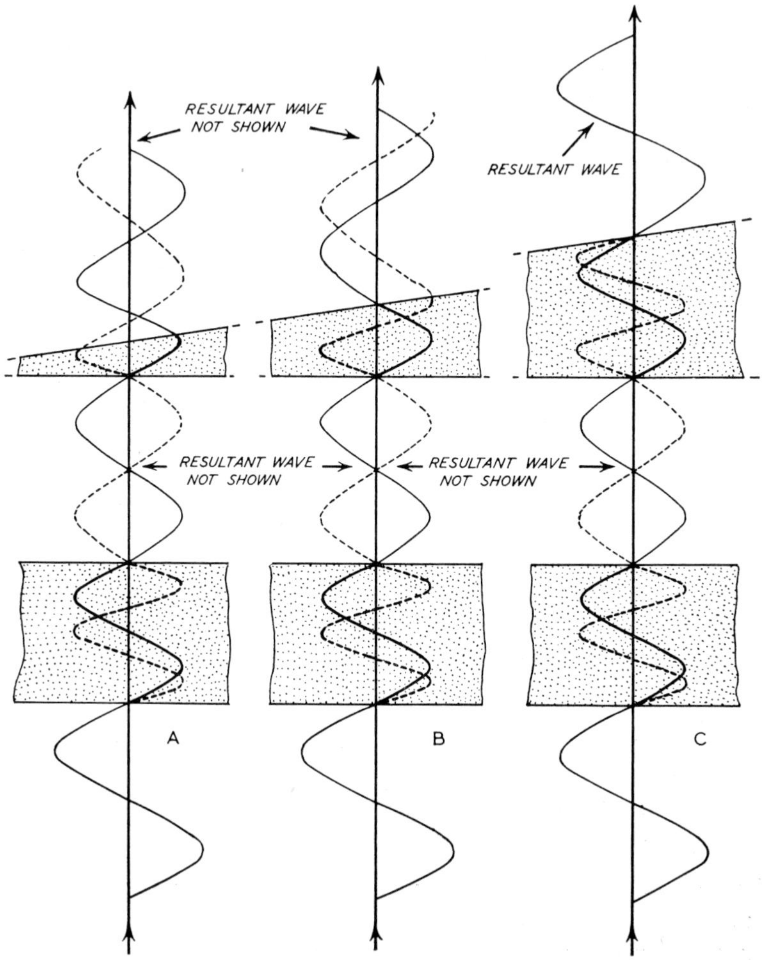

Fig. 11. Two-dimensional diagram showing compensation by a quartz wedge. Path difference (phasal difference) produced by crystal plate has been nullified by quartz wedge in position *C*.

crystal plate is resolved into two components, one (dashed line) slower than the other (solid line). Actually the waves corresponding to these components vibrate in mutually perpendicular planes, but are shown

in the same plane for purposes of pictorial representation. The crystal plate has an orientation, thickness, and birefringence such that a path difference (phasal difference) of one-half wave length is produced. The resultant of the waves emerging from the crystal plate is not shown because, in any event, the resultant is again resolved in the quartz wedge. If the trace of the vibration plane of the fast component of the wedge is parallel to the trace of the vibration plane of the slow component of the crystal plate, compensation results when the thickness of the wedge is such that the path difference produced by the crystal plate is nullified, as in Fig. 11*C*.

Abnormal Interference Colors. Abnormal interference colors are produced by uniaxial crystals in two ways. If a crystal is isotropic for a particular color but not for the other colors of the spectrum, the color for which the crystal is isotropic is removed from white light passing through the crystal and, in general, a complementary color appears. Melilite, for example, is sometimes isotropic or nearly so for yellow light and appears blue even though the path difference produced by the mineral is only a fraction of one-half wave length for blue. This blue is called *abnormal blue.*

Minerals which are inherently colored, owing to the differential absorption of the components of white light, give interference colors which depend not only on the path difference produced by a crystal fragment but also on the nature of the light absorbed by the fragment.

Pleochroism (Dichroism). Certain nonopaque crystals absorb light differently in different directions. If tourmaline, for example, is examined under the petrographic microscope with the upper polarizing prism removed, it will change color on rotation of the microscope stage (Fig. 12). When the optic axis of the tourmaline is parallel to the plane of polarization of the lower prism, the light is not as strongly absorbed as when the optic axis is at right angles to this plane. The absorption is expressed by the formula: E, weak; O, very strong.

The variation in color resulting from differential absorption is called *pleochroism* (or *dichroism*) and is described by a *pleochroic formula.* Tourmaline might have the following formula: O = buff; E = neutral gray.

Advantage is taken of the differential absorption of light by tourmaline in the construction of tourmaline tongs (Fig. 13). These are constructed so that two plates of tourmaline cut parallel to the c crystallographic axis may be rotated into any position with respect to each other. If the plates are parallel, light is transmitted through both, but, if the plates are crossed, little or no light passes through.

Extinction Angles. Extinction angles are the angles between extinction positions and a plane, line, or edge in a crystal or crystal fragment.

If a crystal face is used to measure an extinction angle, care must be taken to identify the face, if the angle is to have any significance. For example, consider a quartz crystal lying on one of its prism faces. If the extinction angle is measured from the trace of a prism face, the

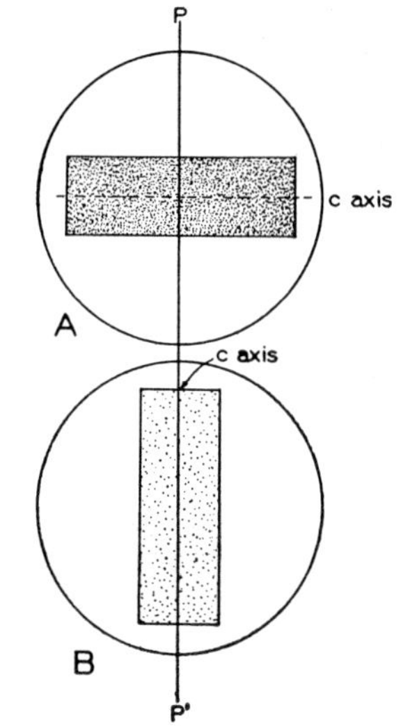

FIG. 12. Tourmaline in plane-polarized light.

A. Optic axis is normal to plane of polarization of lower polarizing prism. Absorption strong.

B. Optic axis parallel to plane of polarization of lower polarizing prism. Absorption weak.

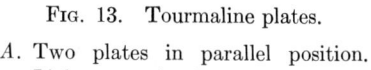

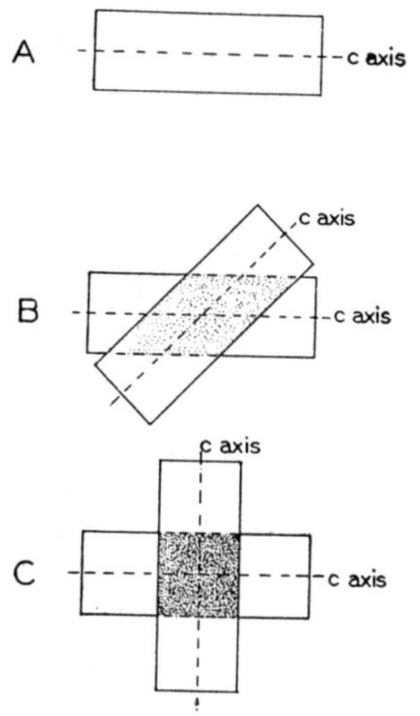

FIG. 13. Tourmaline plates.

A. Two plates in parallel position. Light transmitted easily.

B. Two plates in 45-degree position. Intensity of transmitted light diminished.

C. Two plates at right angles. Transmitted light is practically zero.

angle will be zero, but if a rhombohedron face is used, a much different result is obtained. If cleavage surfaces are used to measure extinction angles, the crystallographic directions of the cleavages should be known. Twin planes, inclusions, etc., may be used if their crystallographic orientations are known.

Index Determination in Uniaxial Crystals. Determination of n_O and n_E in uniaxial crystals is easily made by the immersion method.

The procedure for uniaxial substances is essentially as follows: crystals or powdered fragments are immersed in a suitable index medium on a glass slide. The relief is noted and the indices of the

fragment relative to the immersion medium is determined by using central illumination or oblique illumination. Next, several grains are placed in a medium which more nearly matches the indices of the substance being tested, and, if the match is close, the upper polarizing prism is inserted, and the grains are observed between crossed polarizing prisms.

To measure n_O select a grain which appears black or dark gray during a complete rotation of the microscope stage. In crystals of moderate or high birefringence such a grain lies so that its optic

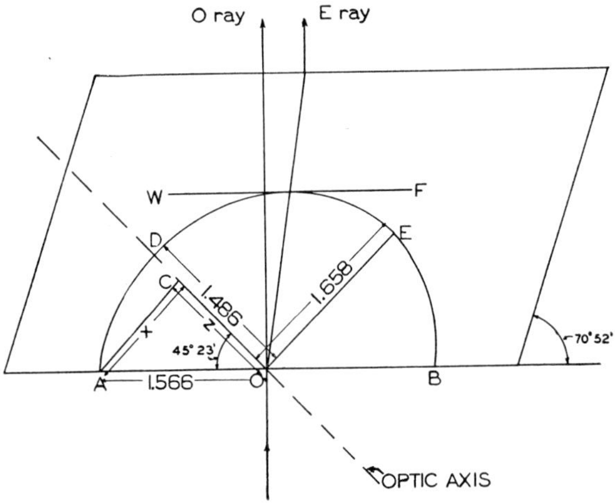

FIG. 14. Relationships of n_O and n_E to crystal directions in calcite.

axis parallels the axis of the microscope. Remove the upper prism, and compare the index of the grain with the immersion medium. If necessary, make further immersions until n_O is matched.

Now, when the n_O index is matched, insert the upper prism and search for a grain with a maximum interference color. Take into account the variation of the interference color with grain size. Grains showing the maximum interference color lie with the optic axis parallel to the microscope stage and, if rotated, extinguish when the n_O and n_E directions coincide with the planes of polarization of the polarizing prisms. Such grains will give both n_O and n_E. The grain is rotated to extinction, and the upper prism is removed. If the index is still matched by the immersion medium, it is known that n_O parallels the plane of the lower polarizing prism, and the grain should be rotated 90 degrees, and it is ascertained whether n_E is greater or less than n_O. By successive immersions n_E is measured in fragments showing maximum colors.

If n_E is higher than n_O, the crystal is positive. If n_E is less than n_O, the crystal is negative.

Any fragment, no matter what its orientation may be, permits determination of n_O, but the full value of n_E is obtained only when the optic axis of a crystal is perpendicular to the axis of the microscope. Certain uniaxial crystals have conspicuous cleavages, and most of the fragments tend to lie on cleavage faces. In some substances it may be necessary to use a very viscous immersion medium, or the problem may be solved by adding powdered cover glass to the immersion oil. This will serve to support the grains in any desired position.

In extreme examples, it may be necessary to resort to calculation to find the full value of n_E. As a matter of fact, in many crystals it is a simple matter to measure $n_{E'}$, the index for the extraordinary wave, for a grain lying on a cleavage not parallel to the optic axis.

Calcite, because of its perfect rhombohedral cleavage, almost invariably lies on a cleavage face in an immersion medium mount. The cleavage meets the optic axis (c axis) at an angle of 45° 23′ (Fig. 14), and the value of the index $n_{E'}$, as determined in a cleavage piece, lies somewhere between n_O and n_E. Suppose that n_O and $n_{E'}$ are measured and prove to be 1.658 and 1.566, respectively. To solve for n_E, the equation of a principal section of the indicatrix is used. This is

$$\frac{x^2}{n_E} + \frac{z^2}{n_O} = 1$$

From Fig. 14 it is seen that

$$x = AO \sin 45^\circ\, 23' = n_{E'} \sin 45^\circ\, 23'$$

and

$$z = AO \cos 45^\circ\, 23' = n_{E'} \cos 45^\circ\, 23'$$

n_O, $n_{E'}$, sin 45° 23′, and cos 45° 23′ are known. Substitution in the equation for the principal section of the indicatrix yields a value of 1.486 for n_E.

If n_O and n_E are known, it is possible to solve for $n_{E'}$ for a grain in any position if the angular position of the optic axis with respect to the microscope stage is known.

CHAPTER X

UNIAXIAL CRYSTALS IN CONVERGENT POLARIZED LIGHT

Introduction. The petrographic microscope with the polarizing prisms crossed may be converted into a conoscope by inserting an accessory lens of short focal length just below the microscope stage and a Bertrand-Amici lens between the upper polarizing prism and ocular. This arrangement of lenses, when used with a high-power objective, permits the observation of interference figures and other interference phenomena. The condensing lens produces strongly converging light which comes to a focus in the plane of the object being examined on the stage of the microscope. This light may be thought of as forming a solid cone of illumination just above the accessory condensing lens.

If the ocular and the Bertrand-Amici lens are removed, interference figures are seen just as clearly but at reduced magnification. The optical system of the conoscope is designed to permit the examination of interference phenomena in a focal plane of the objective, where a real image is produced.

Figure 1 shows the essential components of the optical system of a conoscope. Strongly converging plane-polarized light passing through a crystal plate is brought to a focus above the analyzing prism. Insertion of the Bertrand-Amici lens at the level of the focal plane of the objective lens system converts the microscope into a telescope focused at infinity. Objects such as trees and telephone poles are clearly seen in the conoscope if they stand in the path of a source of daylight illumination.

Interference Figures. Anisotropic crystals under the conoscope yield interference figures consisting of *isogyres* and *isochromatic curves.* The isogyres are black or gray areas which may or may not change position as the microscope stage is rotated. The isochromatic curves are color bands or areas which are systematically distributed with respect to the isogyres.

Interference figures are useful in evaluation of the optic sign and assist in the determination of the optical orientation of a crystal fragment. Moreover, interference figures permit estimation of birefringence and thickness of crystal plates or fragments.

If light moving along parallel or subparallel rays passes through a crystal plate of uniform thickness between crossed polarizing prisms, a uniform interference color results which, as indicated in Chapter IX, depends on the thickness, orientation, and birefringence of the plate. If the orientation is changed and the other factors remain constant, the interference color changes. In the conoscope the orientation of the crystal plate is in effect different for light transmitted along each ray. The interference effects depend on thickness and birefringence as above, but vary in response to a variation in the angles of incidence of the light entering the crystal plate from the substage condensing lens.

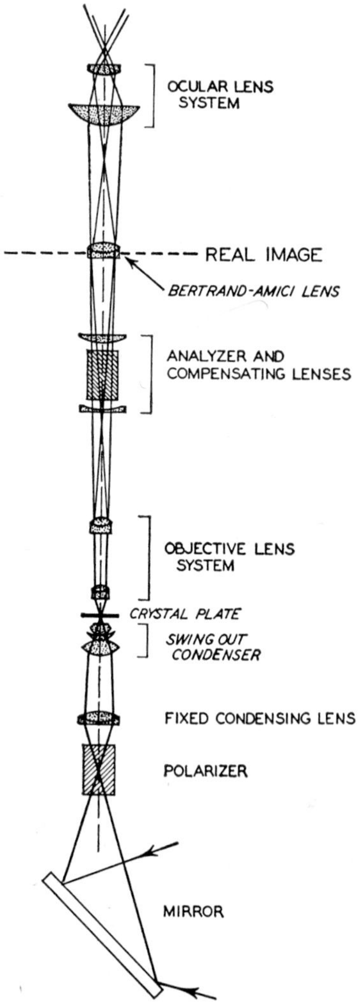

FIG. 1. Optical system of a conoscope.

The effect of variation of the angle of incidence of monochromatic light entering a crystal plate is diagrammatically shown in Fig. 2. The crystal plate in the diagram is cut normal to the optic axis of a positive mineral, as is indicated by the orientation of the ray velocity surfaces. The plane of the drawing lies in a principal section. Light (Fig. 2*A*) traveling a path *AA′* parallel to the optic axis suffers no double refraction and passes through the crystal as if it were isotropic. However (Fig. 2*B*), light moving along each of the parallel rays *AB* and *A′B′*, upon entering the crystal, is doubly refracted, and is resolved into two components vibrating in mutually perpendicular planes. One component, the extraordinary or *E* component, vibrates in the principal section; the other component, the ordinary or *O* component, vibrates normal to the principal section. If, as in Fig. 2, a crystal is positive, the *O* component moves with greater velocity than the *E* component. The velocities are indicated by the

spacing of the arrows and the dots. If rays AB and $A'B'$ are properly spaced, components traveling along BC' and $B'C'$ are refracted on leaving the crystal so as to travel the same path $C'D'$ and, upon passing through the analyzing prism, produce an interference effect depending on the wave length of the light and the path difference.

As the inclination of the incident light increases, the path difference produced by the plate increases as suggested in Fig. 2C. That is, the light in rays in the outer layers of the cone of light above the con-

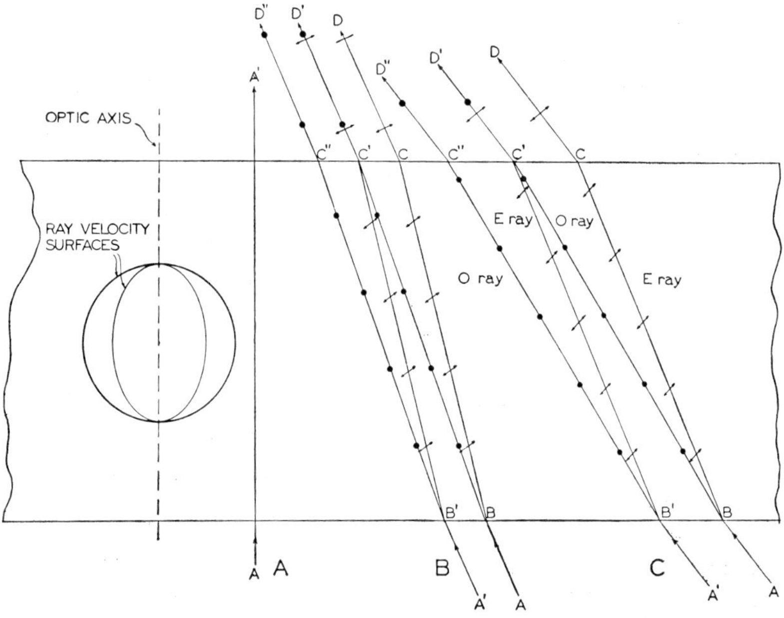

FIG. 2. Double refraction of light with different angles of incidence.

densing lens produces interference effects higher in the spectrum than the light traveling more nearly parallel to the optic axis of the crystal.

Optic-Axis Figures. The effect of a uniaxial crystal plate on convergent light is further illustrated in Fig. 3, which shows a cone of light entering a crystal plate so as to be symmetrically disposed with respect to the optic axis. Suppose that the light is monochromatic. Then, depending upon the varying angle of incidence of the light in the incident cone, various path differences result. The path difference of the components emerging from the crystal plate increases at the same rate outward along all radial lines passing through the point of emergence of the optic axis. Figure 4 shows the loci of points of emergence of components having $\frac{1}{2}\lambda$, $\frac{3}{2}\lambda$, and $\frac{5}{2}\lambda$ path differ-

ence (dashed lines) and 1λ, 2λ, and 3λ path difference (solid lines). In plan the resulting interference figure consists of alternate concentric circles of light and darkness and a black cross, the arms of

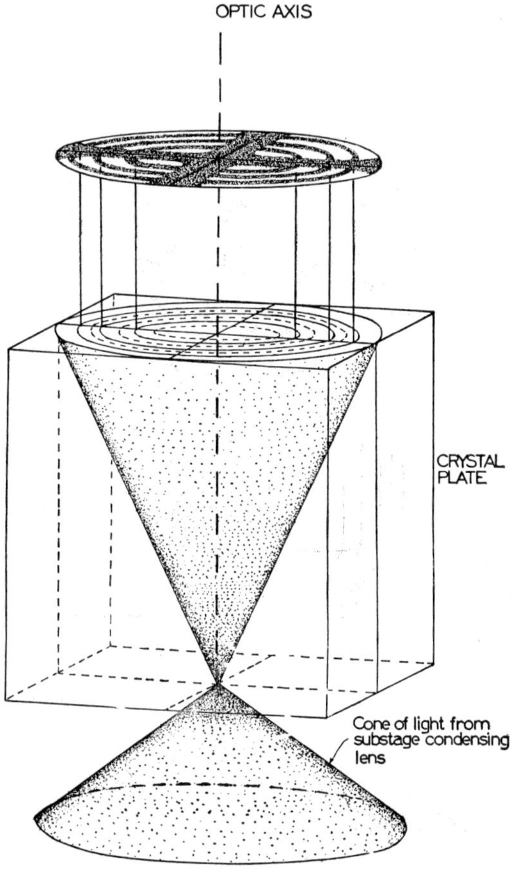

FIG. 3. Effect of uniaxial crystal plate on convergent monochromatic light.

which parallel the planes of polarization of the upper and lower polarizing prisms.

The circles of light form where the components of light emerging from the crystal plate have a path difference of $(n/2)\lambda$, where n is equal to 1, 3, 5, etc. The dark circles (stippled) form where the path difference is $n\lambda$, if n is equal to 1, 2, 3, etc.

The explanation of the black cross and the color rings rests on the use of vector diagrams. Figure 5*A* demonstrates the relations that exist for monochromatic light in an interference figure where the com-

ponents emerging from the crystal plate have a path difference of $n\lambda$ wave lengths, where $n = 1, 2, 3$, etc. It should be remembered that in a symmetrical optic-axis interference figure, the points of emergence of the components of the same path difference lie on circles. Let PP' be the plane of polarization of the polarizing prism, and let Ob equal

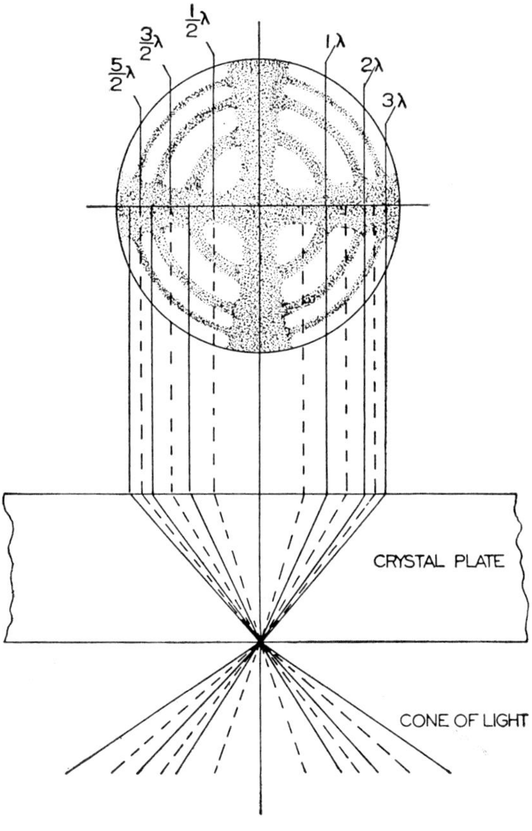

Fig. 4. Uniaxial optic axis interference figure in polarized convergent monochromatic light.

the amplitude of the light vibrating in this plane. If XX' and YY' are the traces of the planes of vibration in the crystal plate, the wave of amplitude Ob upon entering the plate will be resolved into two components having amplitudes Oc and Oc'. Oc and Oc' are obtained by dropping lines perpendicular to OY and OX from b.

If the path difference is $n\lambda$, where n equals a whole number, the component vibrating in the XX' plane will start its motion from O to c at the same instant that the component vibrating in the YY' plane

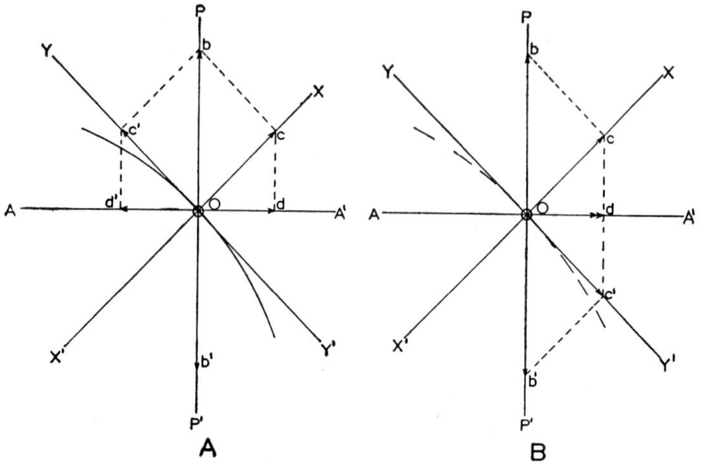

Fig. 5. Explanation of isogyre and color rings in uniaxial optic-axis figure.
A. Path difference = $n\lambda$. *B*. Path difference = $(n/2)\lambda$.

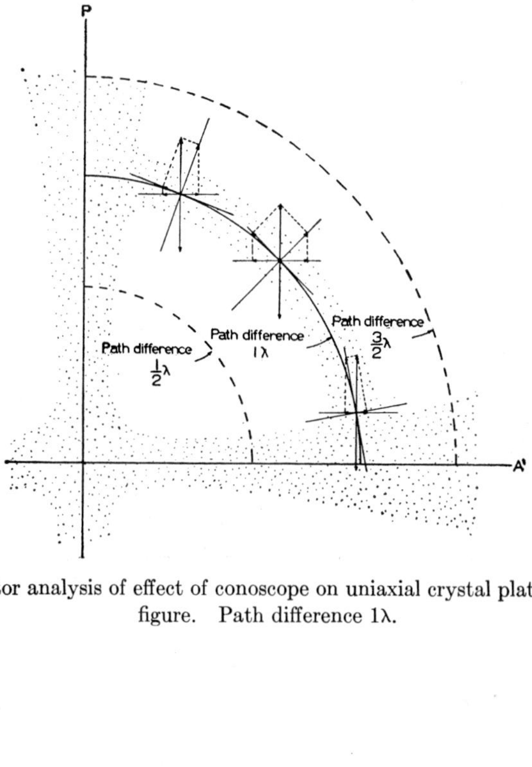

Fig. 6. Vector analysis of effect of conoscope on uniaxial crystal plate. Optic-axis figure. Path difference 1λ.

starts its motion from O to c'. The components of Oc and Oc' which pass through the analyzing prism are obtained by dropping normals on AA' and are, respectively, Od and Od'. From the diagram it is seen that these components are opposite and equal. The result is darkness. In Fig. 6 the indicated points of emergence all lie on a curve of darkness because the path difference is $n\lambda$, but it should be noticed that, as

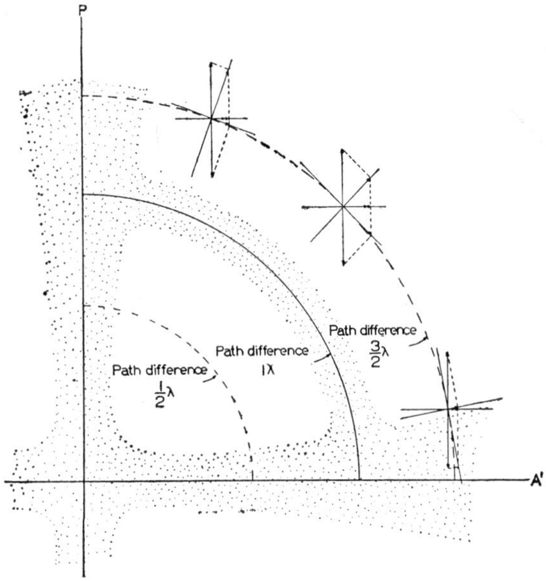

FIG. 7. Vector analysis of effect of conoscope on uniaxial crystal plate. Optic-axis figure. Path difference $\frac{3}{2}\lambda$.

the points of emergence swing away from the 45-degree position, the amplitudes of the opposing components in the analyzing prism decrease.

In Fig. 5*B* the path difference produced by the crystal plate is $(n/2)\lambda$, $n = 1, 3, 5$, etc. As in Fig. 5*A*, OB is the amplitude of a wave of monochromatic light vibrating in the plane of polarization of the polarizing prism PP'. XX' and YY' are the traces of the planes of vibration of the crystal. If the path difference is $(n/2)\lambda$, the component vibrating in the XX' plane starts its motion from O to c at the same instant that the component vibrating in the YY' plane starts its motion from O to c'. The components of Oc and Oc' passing through the analyzing prism are obtained by dropping normals on AA'; both components equal Od, move in the same direction, and result in light of amplitude twice Od. In Fig. 7, as the planes of

polarization of the light emerging from the crystal plate approach coincidence with the planes of polarization of the upper and lower prisms, the amplitudes *Od* approach zero. At coincidence the amplitudes are zero and no light leaves the upper prism, and, as a result a black cross, the isogyre, forms.

Useful in explaining isogyres is the skiodrome illustrated in Fig. 8. A skiodrome is an orthographic projection of *curves of equal velocity* as they would appear on a sphere, supposing that the light source is at

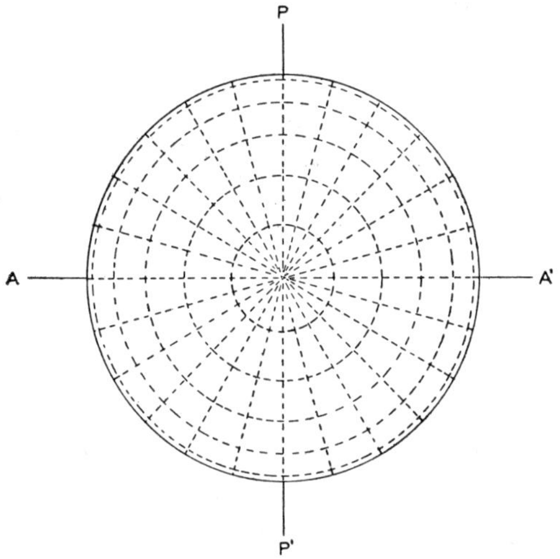

FIG. 8. Skiodrome of uniaxial crystal. Plane of projection normal to optic axis.

the center of the sphere. The plane of projection of the skiodrome shown in Fig. 8 is normal to the optic axis.

The traces of the vibration planes of the two components of light transmitted along any ray at the point of emergence on the skiodrome are parallel respectively to the radial line and to the tangent to the circle passing through the point of emergence. By using this rule Fig. 9 was constructed. Because light vibrating in a plane, the trace of which is parallel or nearly parallel to the upper and lower polarizing prisms is cut out, a black cross will result as indicated by the stippling in Fig. 9. If the optic axis is exactly parallel to the axis of the microscope, the cross maintains a constant position during a 360-degree rotation of the crystal plate.

If a white-light source is used, the uniaxial optic-axis figure in general consists of a black cross and a series of concentric color circles. From the center outward, the colors rise in the spectrum as the path

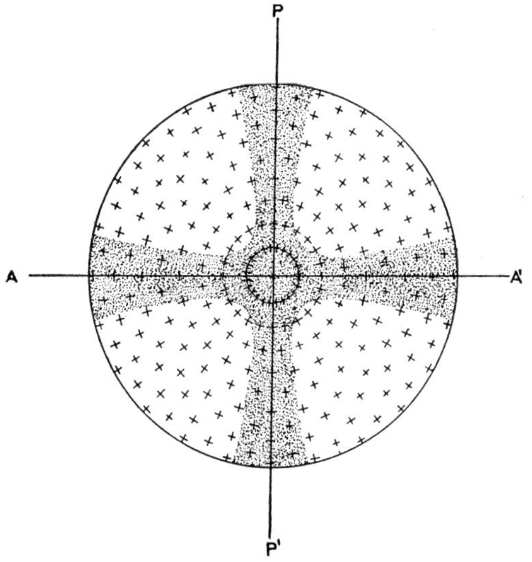

FIG. 9. Isogyre constructed from the skiodrome.

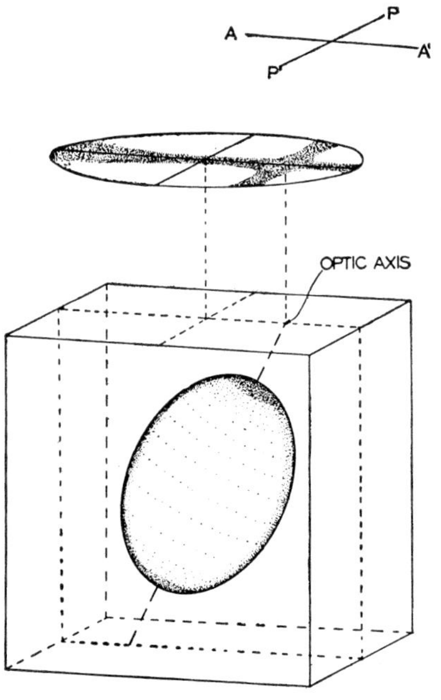

FIG. 10. Off-center optic-axis figure. Point of emergence of optic axis in field of microscope.

difference increases. Gray next to the center of the figure grades outward into yellow, then red, violet, blue, green, yellow, etc. If the birefringence of the crystal plate is low or the plate is very thin, color curves are not present, and only a diffuse isogyre is seen. As the plate increases in thickness or birefringence, or both, the color

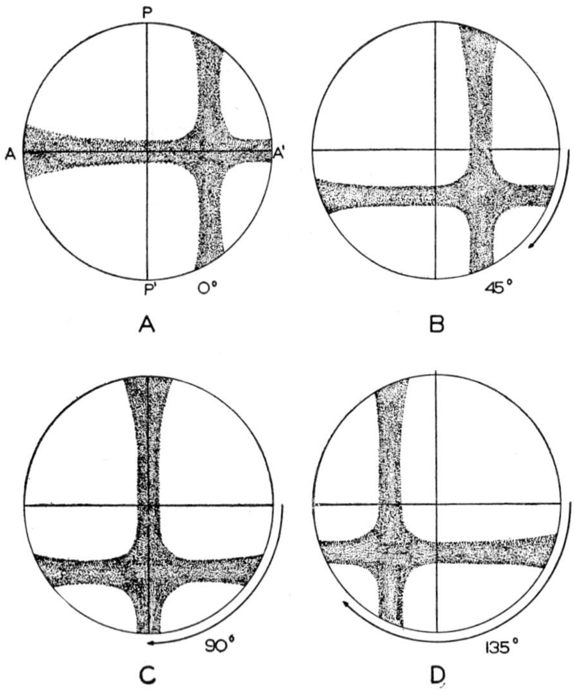

FIG. 11. Clockwise rotation on an off-center optic-axis figure. Point of emergence of optic axis in field of microscope.

bands become more numerous and more closely spaced and the isogyre becomes more sharply defined.

Under the orthoscope, crystal plates cut exactly normal to the optic axis remain black on rotation between crossed polarizing prisms. Accordingly, in the search for a fragment or grain which will give an optic-axis interference figure, the observer should look for a grain which has a black or dark-gray interference color during a complete rotation of the microscope stage.

Off-Center Optic-Axis Figures. Perfectly centered optic-axis figures are rarely observed, and it becomes necessary to work with more or less off-center figures. But such interference figures can be as informative as the more symmetrical ones. Off-center figures result

when the optic axis is not parallel to the axis of the microscope. Two possibilities are considered to illustrate cause and effect in such interference figures.

Figure 10 shows a positive uniaxial crystal plate in which the indicatrix is slightly inclined with respect to the surface of a crystal

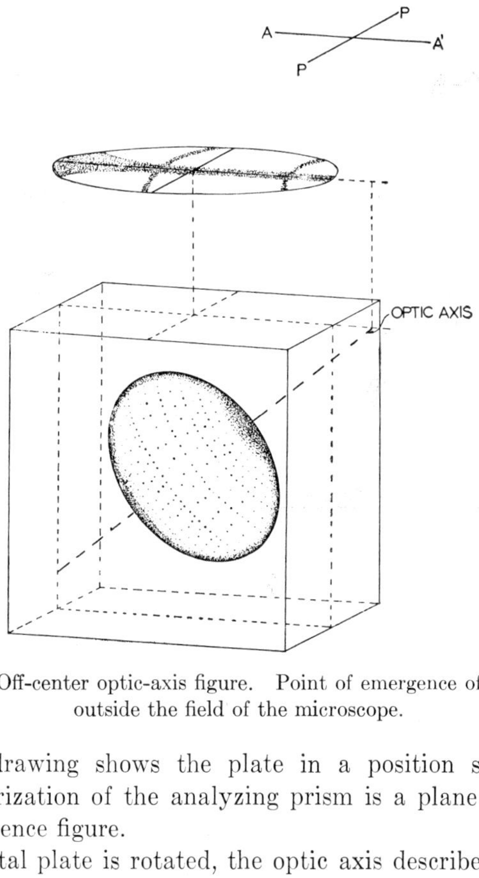

FIG. 12. Off-center optic-axis figure. Point of emergence of optic axis outside the field of the microscope.

plate. The drawing shows the plate in a position such that the plane of polarization of the analyzing prism is a plane of symmetry of the interference figure.

As the crystal plate is rotated, the optic axis describes a cone, and the point of emergence of the optic axis, a circle. Figure 11 shows in plan the effects of rotating the crystal plate. The arms of the isogyres move across the field essentially as straight bars, paralleling the traces of the planes of polarization of the upper and lower polarizing prisms.

Figure 12 illustrates an example in which the point of emergence of the optic axis lies outside the field of the microscope. The crystal plate is uniaxial negative, as indicated by the outlines of the indicatrix, and has a higher birefringence than the example described above.

As the crystal plate is rotated, the point of emergence of the optic axis describes a circle which lies outside the field of view. The effects of rotation are indicated in Fig. 13, which shows that the brushes of the isogyre move as nearly straight bars across the field of the microscope as the stage is rotated. A study of the motion of the bars permits location of the approximate position of emergence of the optic

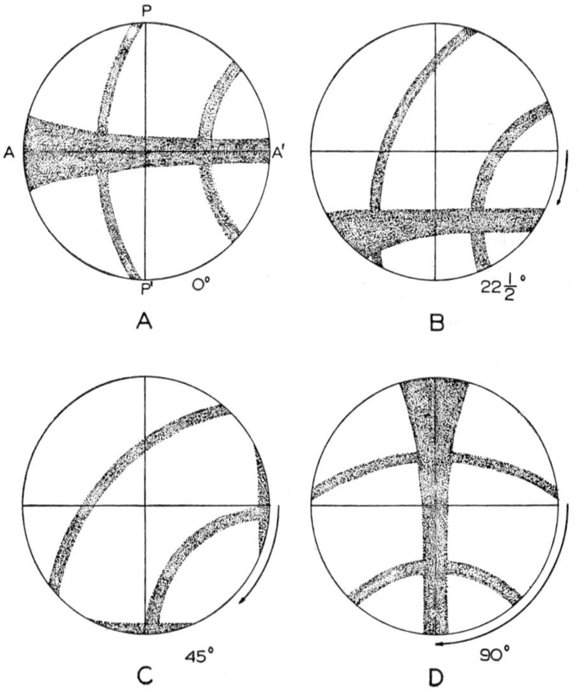

Fig. 13. Clockwise rotations on optic-axis figure. Point of emergence of optic axis outside the field of the microscope.

axis. For example, the isochromatic curves and brushes in Fig. 13*C* suggest that the point of emergence of the optic axis lies in the lower right-hand quadrant.

Flash Figures. Uniaxial crystal plates lying so that the optic axis is parallel to the microscope stage give *flash figures* under the conoscope. In the parallel position, that is, when the optic axis parallels the plane of polarization of either the upper or lower polarizing prism, the flash figure is a poorly defined black cross. The cross separates into two hyperbolae which quickly leave the field upon slight rotation of the microscope stage from the parallel position. Figure 14 illustrates a flash figure in the parallel position. Upon rotation the

hyperbolic segments of the black cross leave the field of the microscope in the quadrants containing the optic axis. Thus the flash figure becomes useful in determining optic sign and orientation.

In the 45-degree position color effects are present which depend upon the thickness and birefringence of the section. In sections of low birefringence or slight thickness (Fig. 15) the color in the center of the field grades outward to a color lower in the spectrum in the quadrants containing the optic axis. In the opposite quadrants the

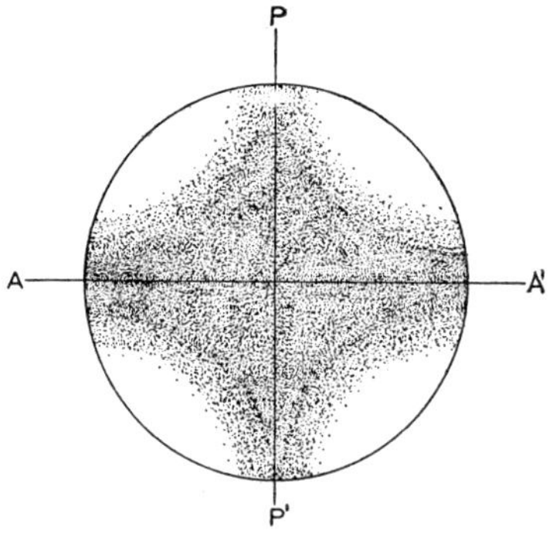

FIG. 14. Flash figure in parallel position.

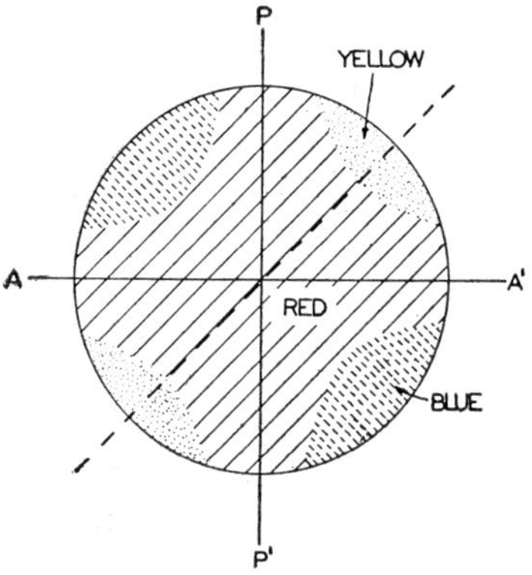

FIG. 15. Distribution of colors in flash figure in 45-degree position.

color rises in the spectrum from the center of the field toward the edge. The distribution of the colors locates the position of the optic axis.

The origin of the black cross is apparent when it is considered that in the parallel position the traces of vibration planes of the crystal plate are parallel to the traces of the planes of polarization of the polarizing prisms. But the color distribution in the figure when the crystal plate is in the 45-degree position is not so easily understood.

Figure 16 is a principal section of a crystal plate showing doubly refracted light and the ray velocity surfaces. Light following a ray (*A*) normal to the optic axis is resolved into two components, the velocities of which are respectively at a maximum and minimum. In white light an interference color appears in the center of the microscope field; the particular color is a function of the path difference produced by the plate. Now consider light moving along the rays (*B*) slightly inclined to the crystal plate. Although the paths of light in these rays are longer than in *A*, the path difference is less, and a lower interference color results. This explains the gradation

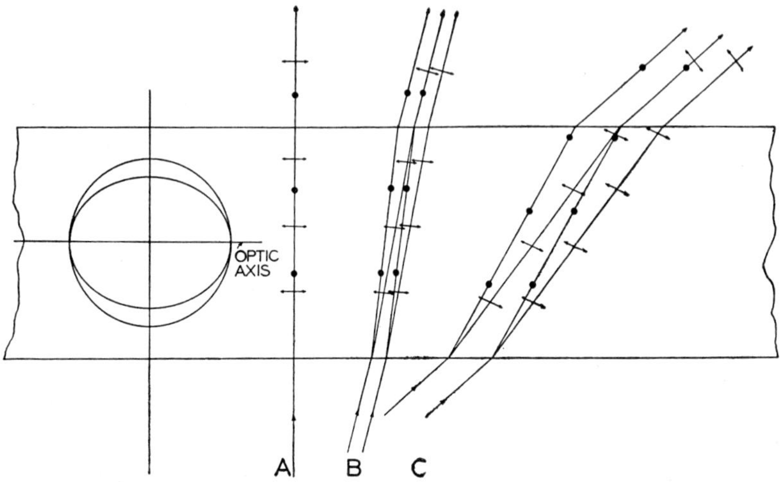

Fig. 16. Explanation of flash figure in 45-degree position. Section parallel to optic axis.

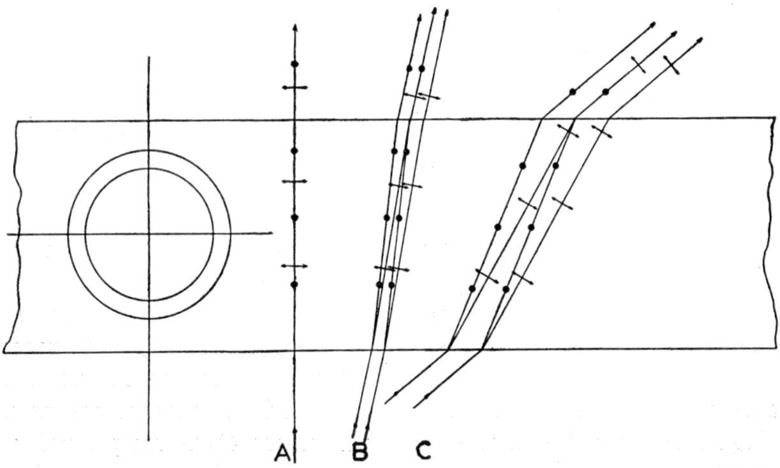

Fig. 17. Explanation of flash figure in 45-degree position. Section normal to optic axis.

to lower colors in the quadrants containing the optic axis. However, in *C* the path difference produced by the plate is greater than in *A* because the greater distances traveled by light along its rays more than offset the decrease in differential velocity. Accordingly, in plates of high birefringence or great thickness the color at first goes down in the spectrum and then, at some distance from the center, rises again.

The rise of color in the quadrants not containing the optic axis is explained by reference to Fig. 17, which shows a section normal to

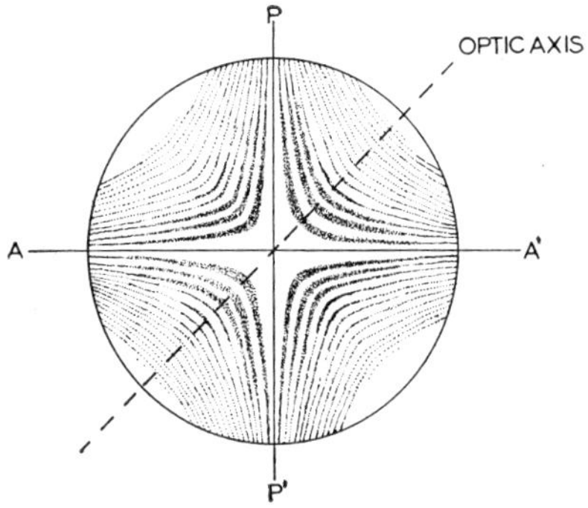

FIG. 18. Interference figure of highly birefringent crystal plate cut parallel to optic axis. Monochromatic light.

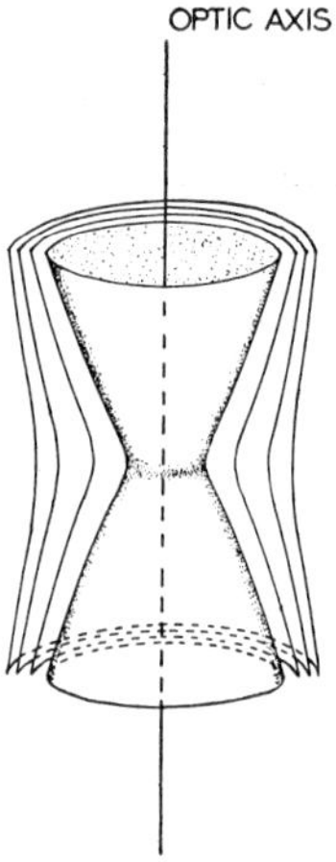

FIG. 19. Nested uniaxial Bertin's surfaces sectioned parallel to optic axis.

the optic axis. The ray velocity surfaces are shown in cross section and consist of two concentric circles. The differential velocity of the components of light moving along all rays in the crystal in the plane of the drawing is the same. Accordingly, the path difference increases with the inclination of the rays, and in the interference figure the color rises from the center toward the edge of the field.

Figure 18 shows the interference figure in the 45-degree position of a highly birefringent crystal plate cut parallel to the optic axis. The light source is monochromatic. The figure consists of an illuminated center and a series of nearly hyperbolic alternating curves of light and darkness. In white light the dark and light curves are replaced by isochromatic curves of all the colors of the spectrum. These grade outward into areas showing white of the higher order.

Further understanding of isochromatic curves results from a con-

sideration of Fig. 19, which shows a series of nested *Bertin's surfaces* for monochromatic light. Bertin's surfaces are *surfaces of equal path difference.* Cross sections are seen in interference figures. The particular cross section of Bertin's surfaces that is seen under the conoscope depends on the orientation, birefringence, and thickness of the crystal plate. For example, a crystal cut normal to the optic axis yields a series of concentric circles. A crystal section parallel to the optic axis gives results like those in Fig. 18. If the surfaces shown in Fig. 19 are surfaces of equal path difference of $(n/2)\lambda$, the cross sections correspond to curves of light in the interference figure. If the surfaces represent equal path differences of $n\lambda$, cross sections correspond to curves of darkness.

Chapter XI

OPTICAL ACCESSORIES

Introduction. Optical accessories have an important and varied use in optical crystallography. Many special types of accessories have been designed to assist in observation and measurement, but relatively few of these are needed for routine work. Three accessories—the quartz wedge, the gypsum plate, and the mica plate—are standard equipment and suffice for all but the most exacting investigations. Present-day manufacturers of microscopes have standardized the construction of the accessories, but it should be kept in mind that older microscopes are often accompanied by accessories of diverse and sometimes unusual construction.

Certain expressions referring to the passage of light through optical accessories are in common use. Light passing through the birefringent crystal plates commonly employed in the construction of optical accessories in general is resolved into two components vibrating in mutually perpendicular planes. Inasmuch as one component travels through the plate with greater velocity than the other, a distinction may be made between the *fast* and *slow components.* One may speak also of the *vibration directions* of light in *fast* and *slow rays, fast* and *slow waves,* or *fast* and *slow components,* if it is understood that the vibration directions to which reference is made are actually the directions of the *traces of the planes* in which the fast and slow components vibrate. The *trace of a vibration plane* is the intersection of the plane with a surface such as a crystal face or the surface of a polarizing prism.

The Quartz Wedge. The quartz wedge is used to determine the directions of the traces of the vibration planes of fast and slow components in crystal plates, to determine the order of interference colors, and to make optical-sign determinations with or without interference figures.

Quartz wedges usually are cut so that the long direction of the wedge is parallel to the fast direction of the quartz, that is, parallel to the direction of the trace of the plane of vibration of the ordinary component. The slow component (extraordinary component) vibrates in a plane normal to the long direction of the wedge.

Some wedges are cut so that the slow component vibrates in a

plane the trace of which is parallel to the long dimension of the wedge. In any event, the wedge should be tested with a crystal of known optical orientation. The theory of the quartz wedge has

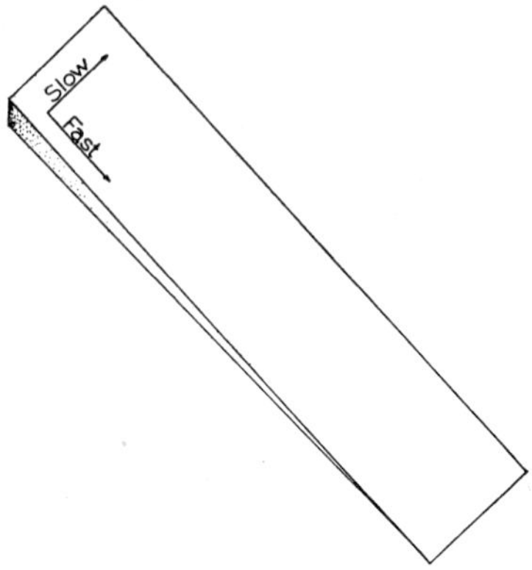

FIG. 1. Quartz wedge.

been described in a previous chapter and will not be discussed here. Figure 1 is a drawing of a typical quartz wedge.

The Gypsum Plate. The gypsum plate is made of clear gypsum or selenite cut or cleaved to such a thickness that it gives a first-order red interference color. Actually, the color is closer to violet than red. The plate is usually mounted in a metal holder so that the trace of the vibration plane of the fast component parallels the longer direction of the holder, and the trace of the vibration plane of the slow component is normal to the longer direction.

The gypsum plate sometimes is called the *sensitive tint* or *red of first-order* plate. In use very slight changes in the interference color of the plate are very apparent to the observer.

The gypsum plate is used for sign determination with crystals or interference figures, for determination of the positions of the traces of vibration planes in crystal plates, and for exact determination of extinction positions. It is most advantageously used with crystals of low birefringence or slight thickness, or both.

The gypsum plate like the quartz wedge usually is inserted into the microscope in the 45-degree position. If, for example, it is desired to determine the direction of the traces of the planes of vibration of the slow and fast components in a crystal plate with a first-order gray interference color, the crystal is turned to extinction and then to the 45-degree position. If the fast component in the crystal plate coincides with the fast component in the gypsum plate, the original violet

color of the plate goes up in the spectrum, say to blue. If the slow component of the crystal is parallel to the fast component of the gypsum plate, the color will go down, perhaps to yellow. It is important to observe that the effect is that of the crystal on the color of the gypsum plate and not the reverse.

The gypsum plate assists in the determination of exact extinction position. A crystal, as it is rotated, approaches extinction gradually, and it is sometimes difficult to ascertain when complete extinction is reached. If the gypsum plate is inserted when the crystal plate is exactly at extinction, the color that is seen will be that of the gypsum plate. If the crystal plate is rotated slightly in either direction, a change in the color of the gypsum plate will be immediately apparent.

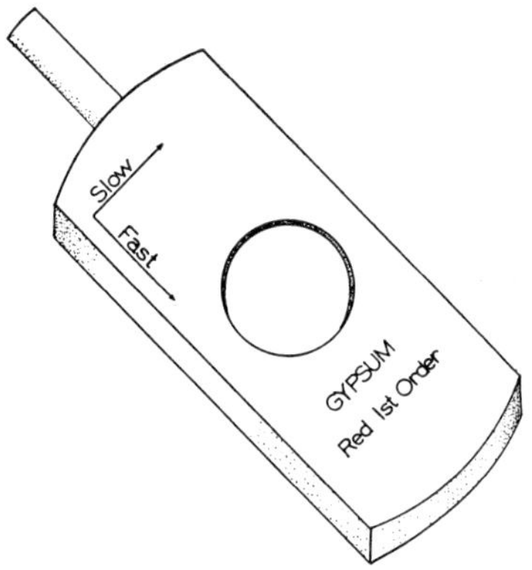

FIG. 2. Gypsum plate.

FIG. 3. Mica plate.

If the crystal does not fill the entire field of view, it may be rotated until the color over the crystal exactly matches the color of the gypsum plate alone. The crystal is then at extinction.

Figure 2 shows a sketch of a common type of gypsum plate.

The Mica Plate. The mica plate, also called the *quarter undulation plate,* or $\frac{1}{4}\lambda$ *plate,* is made of a sheet of muscovite mica cleaved so that its interference color in white light is pale neutral gray. A plate of proper thickness produces a path difference of one-fourth wave length for sodium light.

The mica plate is used to determine optical sign from interference figures and may be used to ascertain the directions of fast and slow components in crystal plates or fragments. The mica plate, when superimposed over a crystal in the 45-degree position in effect increases or reduces the path difference produced in the crystal by the equivalent of one-fourth wave length for sodium light.

The mica plate is usually mounted in a metal holder and, like the gypsum plate, is ordinarily oriented so that the trace of the plane of vibration of its fast component is parallel to the long direction of the holder. It is most useful if used with crystals showing interference colors above white of the first order. Suppose, for example, that the mica plate is superimposed over a crystal plate with a second-order green interference color. If the fast components in crystal and mica plate are parallel, the total path difference is increased, and the color goes up the scale, perhaps to yellow. If the fast and slow components are parallel, the interference color goes down, say to blue.

Figure 3 shows a sketch of a mica plate as commonly constructed.

Accessories for Measurement of Extinction Angles. For accurate measurement of extinction angles, two accessories have proved very

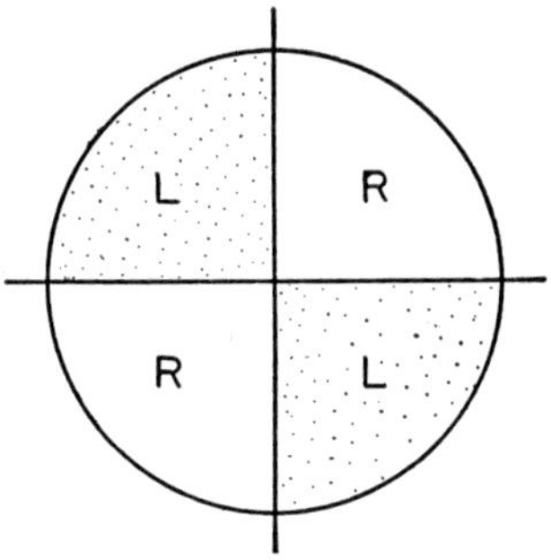

FIG. 4. Bertrand ocular.

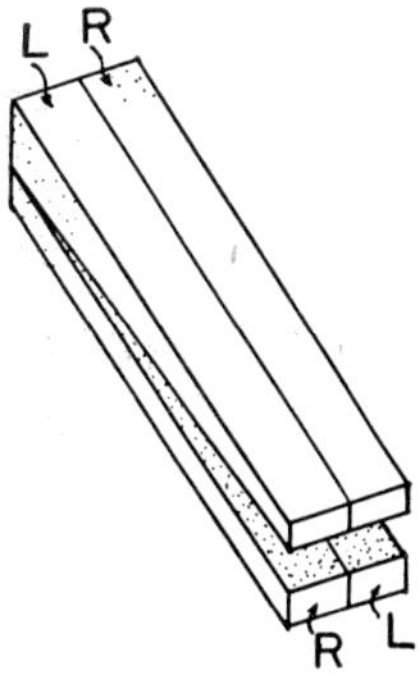

FIG. 5. Wright's biquartz wedge.

useful. These are the *Bertrand ocular* and *Wright's biquartz* wedge. Both require a cap analyzer that is placed above the accessories and rotated to the proper position.

The *Bertrand ocular* consists of four plates of quartz of the same thickness cut normal to the optic axis. Two plates are cut from right-handed quartz and two from left-handed quartz. Pairs of plates are mounted in alternate quadrants so that their rectangular edges correspond to the cross hairs (Fig. 4). The plane of polarization of light from the polarizer is rotated to the same extent by the left and right segments but in opposite directions. When the plane of polarization of the cap analyzer is exactly at right angles to that of the polarizer, the plates all yield a uniform pale bluish-green color.

A birefringent substance on the microscope stage, if not at extinction, causes the quadrants containing right-handed quartz to assume one color, the other quadrants a different color. Exactly at extinction all quadrants have the same color.

The biquartz wedge devised by Wright[1] requires not only a cap analyzer but also a special ocular cut to allow for the insertion of the wedge at its focal plane. The wedge is constructed of two plates, over which are placed two wedges. The pieces are all cut at right angles to the optic axis, two from right-handed and two from left-handed quartz, and are arranged as shown in Fig. 5. Zero rotation of the plane of polarization of light is produced where each half wedge has the same thickness as the underlying plate. A black band extending across the wedge marks the position of zero rotation.

When the wedge is placed over a birefringent substance not at extinction, the illumination of the two halves of the wedge is unequal. At exact extinction the half-wedges are equally illuminated.

Courtesy of Bausch and Lomb Optical Co.

FIG. 6. Holder and eyepiece with graduated quartz compensator.

Accessories for Measurement of Path Difference. Path difference (or phasal difference) produced by a birefringent crystal plate may be accurately measured by several devices. Two are widely used: the graduated quartz compensator and the Berek compensator. The graduated quartz compensator requires a special ocular (Fig. 6) and a cap nicol. Compensation produces a dark line across the wedge. The position of this line on a graduated scale etched on the surface of the wedge gives the path difference directly in millionths of a millimeter ($m\mu$).

The Berek compensator (Fig. 7) is used in the tube slit above the objective and, therefore, does not require a cap analyzer and a special ocular. A plate of calcite 0.1-mm thick is cut normal to the optic axis and mounted on a rotating axis in a metal holder. A calibrated drum controlling the rotation permits measurement of the angular position of the calcite plate.

[1] F. E. Wright, *Am. Jour. Sci.*, Vol. 29, p. 415, 1910.

To determine path difference the crystal plate is rotated on the microscope stage so that the trace of the vibration plane of its fast component is parallel to the trace of the vibration plane of slow component in the inclined calcite plate in the compensator. The angle

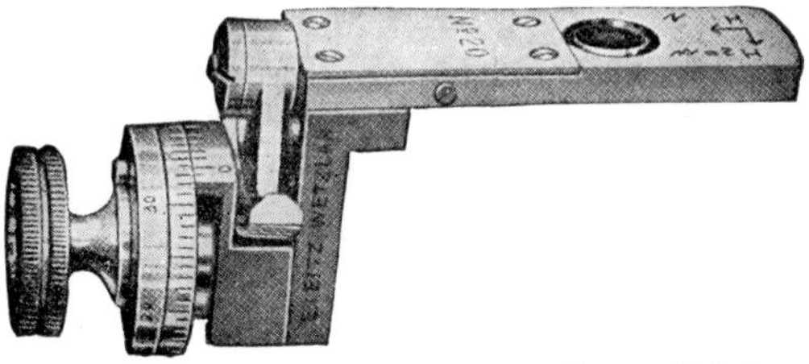

Courtesy E. Leitz and Co.

FIG. 7. Berek compensator.

through which the calcite plate is rotated to reach compensation measures the path difference produced by the crystal plate.

The Berek compensator assists in the determination of optic sign. Also, it may be used effectively to determine the directions of vibration of the fast and slow components in crystals.

The Universal Stage. The universal stage, when attached to the microscope stage, converts the microscope into an instrument of diversified uses. A brief summary of the theory and application of the universal stage is given in Appendix A.

Chapter XII

SIGN DETERMINATION IN UNIAXIAL CRYSTALS

Introduction. By definition uniaxial crystals are positive if the index of refraction for the extraordinary wave, n_E, is greater than the index for the ordinary wave, n_O. If the reverse relationship holds, the crystal is negative. In positive crystals the velocity of the ordinary component is greater than that of the extraordinary component; the reverse is true in negative crystals.

Most methods of obtaining optic sign depend on measurement of the indices of refraction or on the determination of the relative velocities of the ordinary and extraordinary components. Three methods are outstanding: (1) direct measurement of the indices of refraction; (2) determination of direction of the traces of vibration planes of fast and slow components in crystals or cleavage fragments in which the orientation of the *c* crystallographic axis is known; and (3) examination of interference figures, including optic axis and flash figures.

Sign Determination with Indices of Refraction. This method is useful when index determinations are being made in immersion media. The index for the ordinary wave may be measured in all uniaxial crystal fragments. A crystal fragment lying so that its optic axis is parallel to the axis of the microscope produces no path difference for transmitted light and remains black or gray between crossed polarizing prisms during a complete rotation of the microscope stage. A fragment in this position yields n_O only.

A fragment lying so that its optic axis is parallel to the stage of the microscope produces a maximum path difference between the components of transmitted light and, accordingly, shows a maximum interference color. A fragment in this position permits measurement of both n_O and n_E. A grain in an intermediate position gives the ordinary index n_O but will not yield the full value for the index for the extraordinary wave. That is, for grains in random positions n_O remains constant, but the index for the extraordinary component varies with the orientation of the grain.

If it is ascertained that the varying index is consistently higher than the index of the ordinary wave, it is known that the crystal is posi-

tive. Conversely, if the varying index is less than n_O, the crystal is negative.

Sign Determination with Crystals or Cleavage Fragments. Uniaxial crystals and cleavage fragments permit determination of optical sign if they are sufficiently well formed and symmetrical to permit recognition of the direction of the *c* crystallographic axis. The extraordinary component vibrates in the principal section, which includes the optic axis and the *c* crystal axis, and the ordinary component vibrates normal thereto.

Suppose that, as in Fig. 1, two hexagonal crystals lie 45 degrees from extinction and show second-order green interference colors between crossed polarizing prisms. Crystal *A* is oriented so that its optic axis parallels the fast direction of an accessory plate. Crystal *B*,

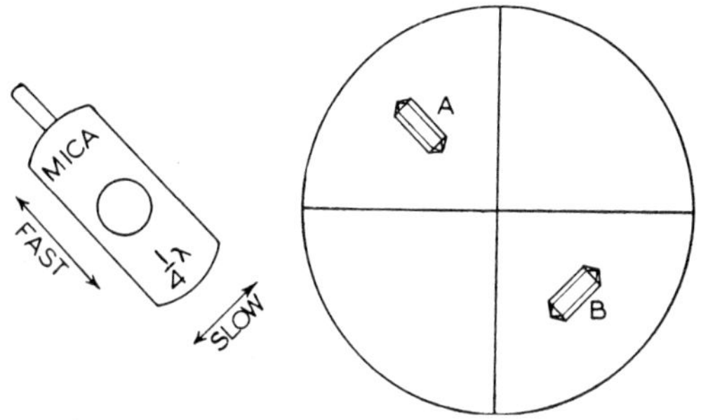

FIG. 1. Sign determination in uniaxial crystals with mica plate.

at right angles to crystal *A*, is oriented with its optic axis perpendicular to the fast direction of the plate. If a mica plate is inserted, and the interference color in *A* goes up in the spectrum to second-order yellow and the color in *B* goes down to second-order blue, the crystals are negative. The color goes up in *A* because the fast direction of the plate parallels the fast component (the extraordinary component) of the crystal. In crystal *B* fast and slow components are parallel.

If a quartz wedge is used, the interference colors go up in crystal *A* and down in crystal *B* as the wedge is inserted. When the path difference of the wedge equals the path difference produced by crystal *B*, the compensation point will have been reached, and the crystal appears gray or black.

With positive crystals the effects on crystals *A* and *B* are the reverse of those described above.

In Fig. 2 two tetragonal crystals lie in the 45-degree position. Suppose that the interference color is first-order gray. The crystals bring about color changes in a gypsum plate and, if the crystals

are positive, crystal *A* causes the first-order red of the plate to fall, perhaps to yellow. Crystal *B* might change the color in the plate to blue. Both color changes are expressions of the fact that in uniaxial positive crystals the extraordinary component is slower than the ordinary component.

Negative crystals under the gypsum plate give effects which are opposite to those described above.

Cleavage fragments may be characteristic enough in certain substances to permit determinations of optical character. In practice, the trace of the vibration plane of the extraordinary component is turned to coincide with the fast direction of the accessory plate to ascertain whether the extraordinary component is faster or slower than the ordinary component. It should be remembered that the extraordinary

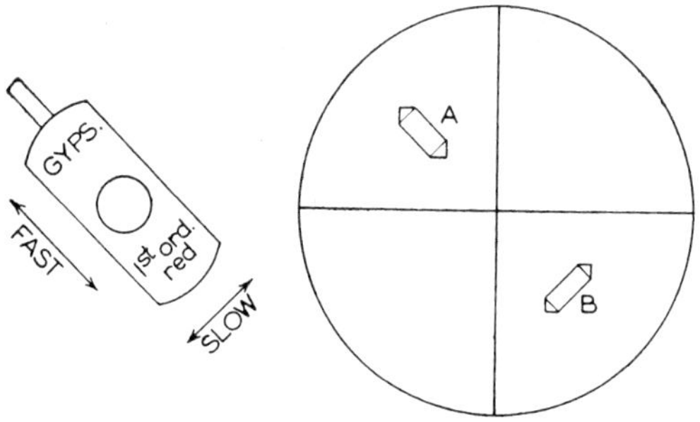

FIG. 2. Sign determination in uniaxial crystals with gypsum plate.

component vibrates in the principal section, which includes both the optic axis and the *c* crystal axis.

In general, in sign determination with crystals, the gypsum plate should be used for crystals with gray or first-order white interference colors, although it may be used effectively with grains showing higher colors. The mica plate and quartz wedge serve well for grains with colors above first-order yellow.

In thick or highly birefringent crystal fragments the interference colors reach white of the higher order. A quartz wedge should be used, and particular attention should be paid to the motion of the conspicuous red color bands in the wedge edges of the fragments. If the fast direction of the quartz wedge parallels the trace of the vibration plane of the slow component of the crystal, the color bands spread out and move in toward the center of the crystal as the quartz wedge is inserted. If the fast directions are parallel, the color bands move toward the edge of the crystal and become more closely spaced.

Sign Determination with Optic-Axis Figures. Uniaxial crystals or fragments lying with their optic axes parallel to the axis of the microscope are black or dark gray between crossed polarizing prisms during a complete rotation of the microscope stage and yield optic-axis figures under the conoscope.

Two types of optic-axis interference figures are shown in Figs. 3 and 4. Figure 3 illustrates the diffuse black cross seen in interference figures of crystal plates having low birefringence or slight thickness, or both. No color curves are present. The gypsum plate is the best accessory for sign determination with this type of figure.

Figure 4 is an interference figure of a more highly birefringent or thicker crystal fragment. Concentric color curves are present, and the arms of the isogyre are narrow and sharply defined. Most suitable

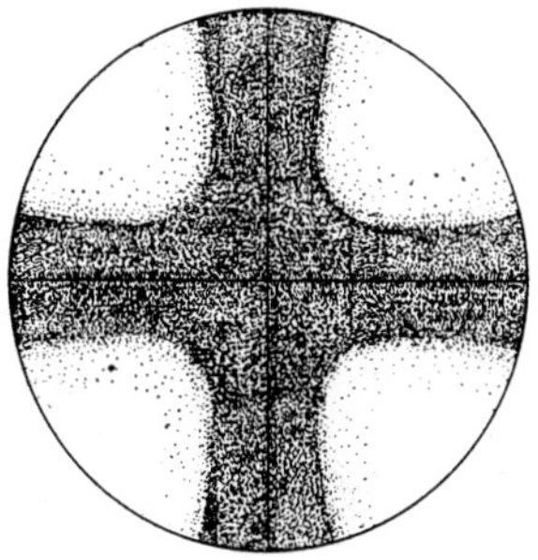

FIG. 3. Diffuse uniaxial optic-axis figure without isochromatic curves.

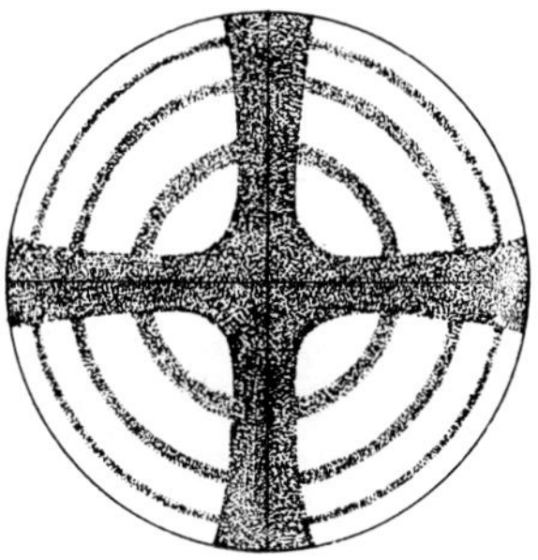

FIG. 4. Uniaxial optic-axis figure showing isochromatic curves.

accessories for determination of sign with this figure are the mica plate and quartz wedge; however, a gypsum plate may be used effectively at times.

The theory underlying the determination of sign in uniaxial optic-axis figures is easily grasped if one keeps in mind the manner of double refraction of the light in the cone entering and passing through the crystal plate. Light transmitted along each ray emerging from the crystal plate consists of two components, one vibrating in the principal section and the other in a plane at right angles thereto. In Fig. 5 light emerging at all points on the dashed circle consists of two components: the extraordinary component E', vibrating in a plane the trace of which is a radial line; and the ordinary component O, in a plane the trace of which is tangent to the circle at the point of emergence. Note that in the 45-degree position the traces of the vibration planes of the extraordinary components in quadrants 2 and 4 are parallel to the vibration direction of the ordinary components in

quadrants 1 and 3. Likewise the ordinary components in quadrants 2 and 4 vibrate parallel to the extraordinary components in quadrants 1 and 3.

Ordinarily, an accessory plate is inserted into the microscope so that its fast direction lies in quadrants 2 and 4 parallel to the traces of the planes of vibration of the extraordinary components. In quadrants 1 and 3 the fast direction of the plate parallels the traces of the vibration planes of the ordinary components. Also, the slow

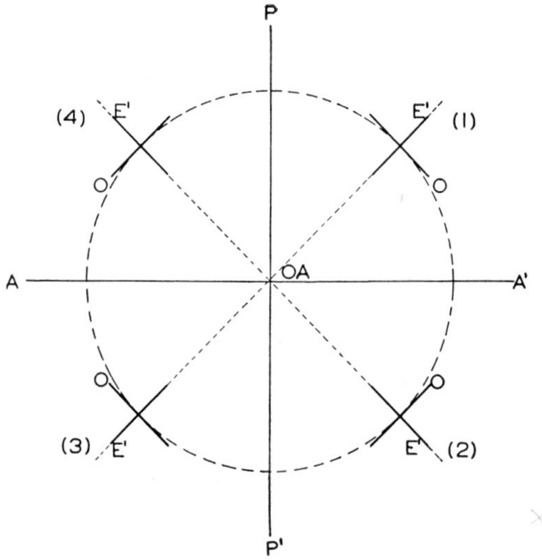

FIG. 5. Traces of planes of vibration of ordinary and extraordinary components in uniaxial optic-axis interference figure.

direction of the accessory plate parallels the traces of the vibration planes of the ordinary components in quadrants 2 and 4 and the traces of the vibration planes of the extraordinary components in quadrants 1 and 3.

In positive crystals the extraordinary component is slower than the ordinary component. In Fig. 6 the effects of the introduction of a gypsum plate on a diffuse optic-axis figure of a positive crystal are shown. In quadrants 2 and 4 the trace of the vibration plane of the extraordinary component is parallel to the fast direction of the plate, and the red of the gypsum plate goes down to yellow. In the opposite quadrants fast components in both the accessory plate and the crystal are parallel, and the color of the plate is elevated to blue.

The opposite effects are obtained by inserting a gypsum plate over an optic-axis figure of a uniaxial negative crystal.

Figure 7 illustrates the action of a mica plate on a positive crystal giving an optic-axis figure showing color curves. Black spots form in quadrants where the fast direction of the plate parallels the trace of the vibration plane of the slow component of the crystal, and the path difference of the light traveling along the rays in the crystal is

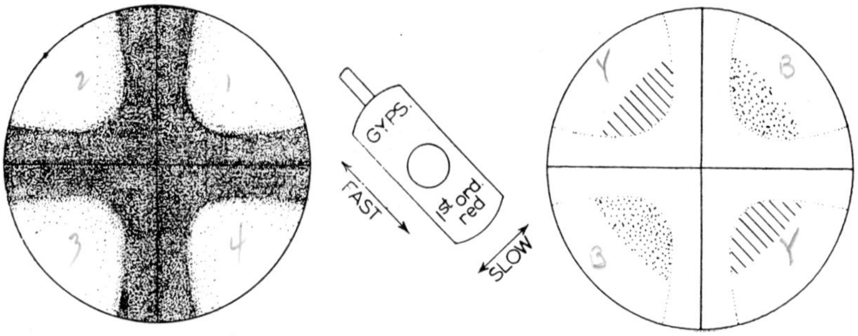

FIG. 6. Determination of sign with gypsum plate. Positive uniaxial optic-axis figure. Stippled area, blue; hatched area, yellow.

exactly compensated for by the $\frac{1}{4}\lambda$ path difference produced by the mica plate. In the quadrants containing the black spots, the isochromatic curves are displaced outward because in these quadrants the mica plate in effect decreases the path difference produced by the

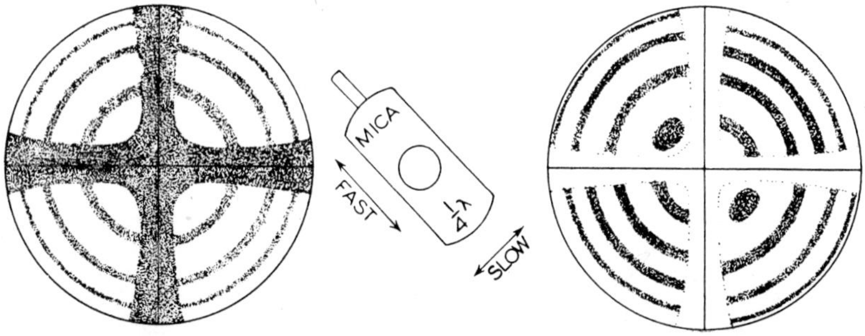

FIG. 7. Determination of sign with mica plate. Positive uniaxial optic-axis figure.

crystal. In the opposite quadrants, where the fast directions of plate and crystal are parallel, the color curves move in because of the $\frac{1}{4}\lambda$ increase of the path difference of all components emerging in these quadrants.

Figure 8 shows the direction of movement of the isochromatic curves in an interference figure when a quartz wedge is pushed into the microscope. Black spots appear where compensation occurs, but the positions of these spots change as the wedge is inserted.

If a gypsum plate is used on interference figures with isochromatic curves, the color changes are seen in the plate very close to the isogyre. Moreover, black curves appear where the plate compensates the first-order red curves.

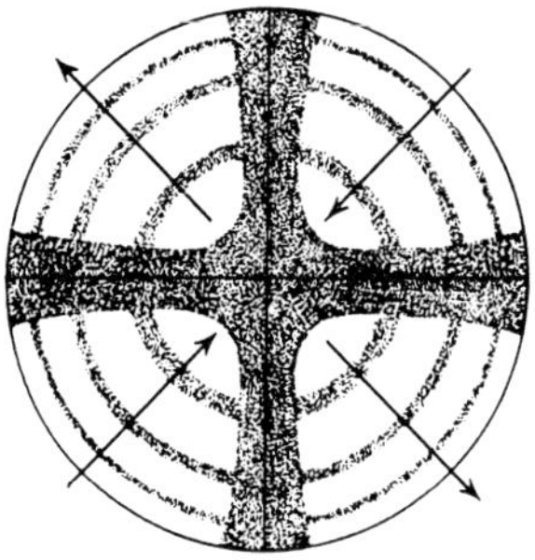

FIG. 8. Movement of isochromatic curves upon insertion of quartz wedge. Uniaxial positive interference figure.

Perfectly centered optic-axis figures are not seen as frequently as off-center figures. However, off-center figures may be used as effectively as the centered ones. Sign is easily determined if the point of emergence of the optic axis lies in the field of view. If, however, the point of emergence of the optic axis lies outside the field of view, its position must be determined by observing the distribution of the isochromatic curves and the motion of the arms of the isogyre during rotation of the stage of the microscope. Figures 9 and 10 show the changes that take place upon insertion of accessory plates over off-center optic-

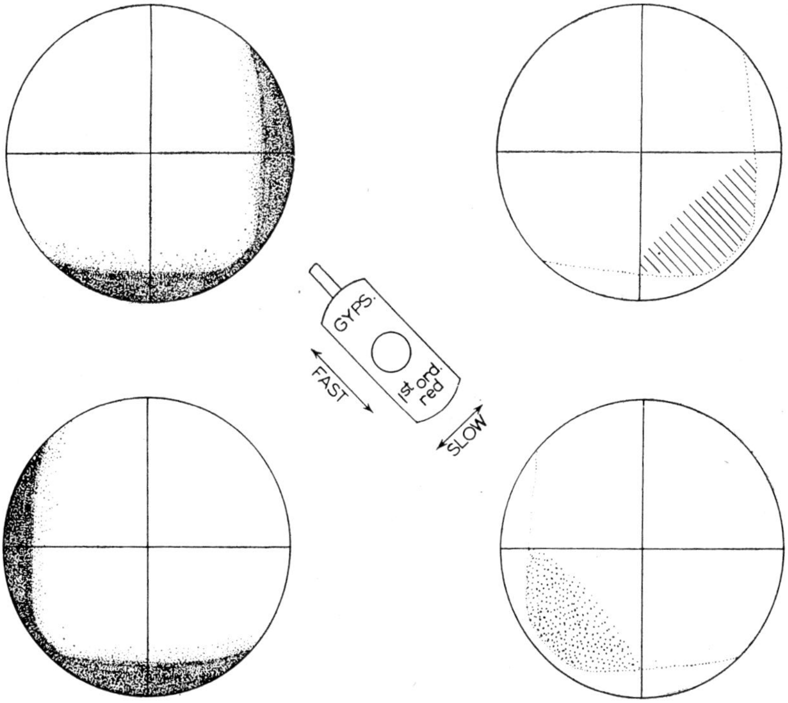

FIG. 9. Positive off-center uniaxial interference figure. Effect of insertion of gypsum plate. Stippled, blue; hatched, yellow.

SIGN OF ELONGATION 141

axis figures, in which the point of emergence of the optic axis lies outside the field of the microscope.

Sign Determination with Flash Figures. If a crystal is known to be uniaxial, a flash figure may be used to determine sign. Upon rotating the microscope stage, the diffuse black cross ordinarily seen in flash figures splits into two hyperbolae which leave the field of

FIG. 10. Positive off-center uniaxial interference figure. Effect of insertion of mica plate.

the microscope in the quadrants containing the *c* crystallographic axis. When the *c* axis is located, the crystal is rotated to the 45-degree position, the Bertrand-Amici lens and the substage condenser are removed, and it is determined by means of accessory plates whether the component vibrating parallel to the *c* axis is faster or slower than that vibrating perpendicular to it.

Sign of Elongation. Hexagonal and tetragonal crystals commonly display a prismatic habit and are elongate parallel to the *c* crystallographic axis. If it is determined in such crystals that the fast component vibrates in a plane the trace of which is parallel or nearly parallel to the long direction of the crystal, the *sign of elongation* is

142   SIGN DETERMINATION IN UNIAXIAL CRYSTALS

said to be *negative,* and the crystal is said to be "length fast." Conversely, if the trace of the vibration plane of the slow component is parallel to the length, the elongation is described as *positive,* and the crystal is said to be "length slow."

However, some crystals are flattened normal to the c axis and in section may give elongate plates. Again, as above, if the fast component vibrates parallel or nearly parallel to the long direction, the sign of elongation is negative; it is positive if the slow component parallels the longer dimension.

Because of the differences of crystal habit, then, the sign of elongation of a crystal may or may not be the same as the optic sign. In the same substance one section may give a positive sign of elongation, and another negative elongation. However, because most of the crystals of a given substance commonly assume the same habit, determination of sign of elongation is useful despite its limitations.

CHAPTER XIII

BIAXIAL CRYSTALS—THE BIAXIAL INDICATRIX

Introduction. Crystals in the orthorhombic, monoclinic, and triclinic systems are *biaxial* and are characterized by three principal refractive indices. Biaxial crystals possess two directions along which monochromatic light moves with essentially the same velocity, as contrasted with uniaxial crystals in which there is only one such direction.

Biaxial Indicatrix. The relations among the indices of refraction in biaxial crystals are best seen in a *biaxial indicatrix* (Figs. 1 and 2), a triaxial ellipsoid which has three planes of symmetry and is so constructed that the three principal indices of refraction of light waves in their *directions of vibration* are equal to its three mutually perpendicular semiaxes. A clear understanding of the function and construction of the ellipsoid is necessary for the development of optical theory for biaxial crystals.

The three principal indices of refraction of waves in their direction of vibration in biaxial crystals are designated as n_X, n_Y, and n_Z and are equal, respectively, to the OX, OY, and OZ semiaxes of the ellipsoid; n_X is the smallest index, n_Y the intermediate index, and n_Z the largest. In a given crystal, the indices of refraction are constant for only one wave length of light. Hence the accompanying diagrams are representative of conditions that exist only when monochromatic light is used. The practical and theoretical implications that arise when white light is used are discussed in a subsequent chapter.

The use of n_X, n_Y, and n_Z to designate the principal indices of refraction is a departure from previous usage. In the 1943 edition of this book, α, β, and γ were employed to indicate the refractive indices. The double use of α, β, and γ to indicate optical constants and, at the same time, the angles between crystallographic axes in the triclinic system is not desirable. Moreover, analytical treatment of the indicatrix in terms of the X, Y, and Z coordinates of conventional geometry seems to favor the suggested change. Equivalent designations are indicated in the accompanying table.

EQUIVALENT DESIGNATIONS OF THE PRINCIPAL REFRACTIVE
INDICES OF BIAXIAL CRYSTALS

n_X	α	N_X	n_α	N_α	X	N_p
n_Y	β	N_Y	n_β	N_β	Y	N_m
n_Z	γ	N_Z	n_γ	N_γ	Z	N_g

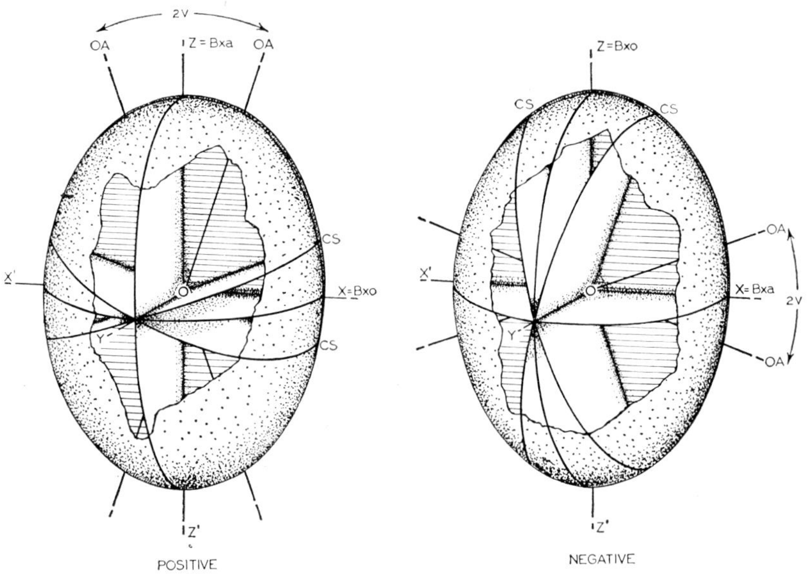

Fig. 1. Biaxial indicatrices. *OA*, optic axis; *CS*, circular section; *Bxa*, acute bisectrix; *Bxo*, obtuse bisectrix.

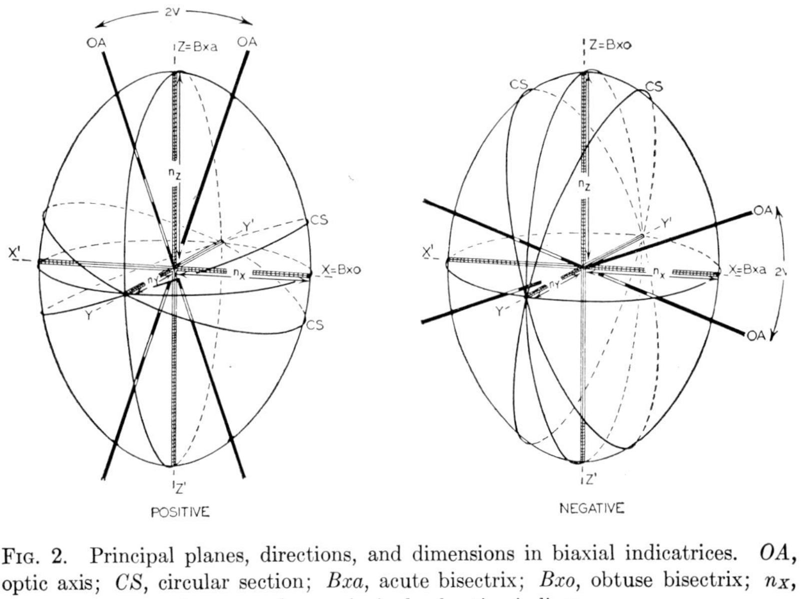

Fig. 2. Principal planes, directions, and dimensions in biaxial indicatrices. *OA*, optic axis; *CS*, circular section; *Bxa*, acute bisectrix; *Bxo*, obtuse bisectrix; n_X, n_Y, and n_Z, principal refractive indices.

In positive biaxial crystals n_Y is closer to n_X than to n_Z. As n_Y approaches n_X, the indicatrix approaches the form of a prolate ellipsoid of rotation—the form of a uniaxial positive indicatrix. In negative biaxial crystals the value of n_Y is closer to n_Z than to n_X, and as n_Y approaches n_Z the indicatrix assumes the form of a negative uniaxial indicatrix lying on its side—in a position such that its optic axis is horizontal.

The *circular sections* of the biaxial indicatrix pass through the Y axis of the indicatrix. All radii of the circular sections are equal to n_Y; thus it is seen that the particular position occupied by a circular section is a function of the value of n_Y as it is related to n_X and n_Z.

The *primary optic axes* are perpendicular to the circular sections. The *optic angle, 2V*, is the smaller angle between the optic axes and lies in the XZ plane, the *optic plane*. In positive crystals the optic angle is bisected by the Z axis of the indicatrix and Z is described as the *acute bisectrix*. In negative crystals X is the acute bisectrix. The axis bisecting the larger angle between the optic axes is designated as the *obtuse bisectrix*.

The Y axis is perpendicular to the plane containing the optic axes (XZ plane) and, accordingly, is called the *optic normal*.

The angle V_Z between the Z axis and an optic axis may be computed from the refractive indices. The following equation is useful and for positive crystals gives an angle less than 45 degrees; for negative crystals, an angle greater than 45 degrees. For negative crystals the computed angle is subtracted from 90 degrees to obtain one-half the optic angle.

$$\tan^2 V_Z = \frac{\dfrac{1}{(n_X)^2} - \dfrac{1}{(n_Y)^2}}{\dfrac{1}{(n_Y)^2} - \dfrac{1}{(n_Z)^2}}$$

The following equations give the angle between an optic axis and the X axis of the indicatrix. In positive crystals the computed angle is greater than 45 degrees.

$$\cos^2 V_X = \frac{(n_Z)^2[(n_Y)^2 - (n_X)^2]}{(n_Y)^2[(n_Z)^2 - (n_X)^2]}$$

or

$$\tan^2 V_X = \frac{(n_X)^2[(n_Z)^2 - (n_Y)^2]}{(n_Z)^2[(n_Y)^2 - (n_X)^2]}$$

When approximations suffice, the following equations are useful:

$$\cos^2 V'_X = \frac{n_Y - n_X}{n_Z - n_X}$$

or

$$\tan^2 V'_X = \frac{n_Z - n_Y}{n_Y - n_X}$$

Figures 3 and 4 show the principal planes of biaxial positive and negative indicatrices. In Fig. 3, $n_X = 1.5$, $n_Y = 1.6$, and $n_Z = 2.0$; in Fig. 4, $n_X = 1.5$, $n_Y = 1.9$, and $n_Z = 2.0$. Note that the relative positions of the circular sections and optic axes in the optic planes are determined by the value of n_Y relative to n_X and n_Z.

The numerical difference between n_X and n_Z gives a measure of the *birefringence* and is geometrically equivalent to the difference between the semimajor and semiminor axes in the XZ plane, the optic plane.

Ray Velocity Surfaces. The triaxial ellipsoid offers an easily visualized means of analyzing the passage of light through crystals. However, it is possible to construct other surfaces which, under certain conditions, are as informative as the indicatrix. One of the most useful concepts is the *ray velocity surface* (ray surface). This is mathematically and geometrically related to the indicatrix and is derivable therefrom.

In Fig. 5, O and O' are point sources of monochromatic light and are coincident with the centers of two identical indicatrices, only one of which is shown. The plane of the drawing includes the optic plane of the indicatrices. Now, let ON' be the direction of movement of a wave generated by the point sources O and O' and vibrating in the plane of the drawing. OT, the dimension of the indicatrix at right angles to the direction of propagation of the wave, is the index of refraction of the wave. The velocity of the wave is proportional to $1/OT$.

To obtain the direction of movement of light along the rays associated with the wave, a line TU is drawn tangent to the indicatrix at T, and a line OS' is drawn through O parallel to TU intersecting the indicatrix at S. OS' is the direction of the rays as determined in this manner. The index of refraction of the light moving along OS' is OM and is determined by drawing a line from O to TU perpendicular to OS'. OM may be thought of as the fraction of OT, the wave index of refraction, giving in effect the refractive index of the light moving in the direction of the ray OS'. Accordingly, OM may be designated as the *ray index of refraction.*

The velocity of light vibrating in the plane of the drawing and traveling along ray OS' is proportional to $1/OM$.

OS and OT are *conjugate radii* and fulfill the condition that radii are conjugate if one radius is parallel to the tangent to the ellipse

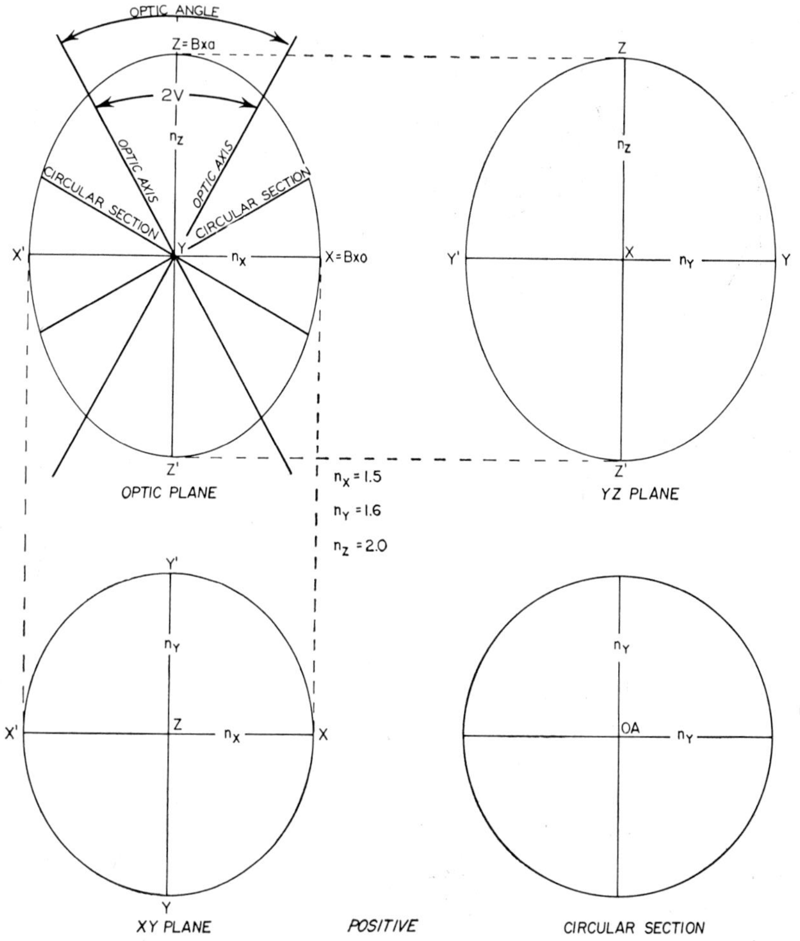

FIG. 3. Principal planes of a positive biaxial indicatrix in which $n_X = 1.5$, $n_Y = 1.6$, and $n_Z = 2.0$.

at the end of the other radius. OT and OS and the tangents to the indicatrix TU and US outline a parallelogram $OTUS$. The area of the parallelogram is equal to the product of the base by the altitude and is $OS \cdot OM$. A theorem of geometry states that the area of all parallelograms enclosed by conjugate radii and the tangents to the

ellipse at the ends of the radii is constant. In Fig. 5, OX and OZ represent a special case in which the conjugate radii are mutually perpendicular.

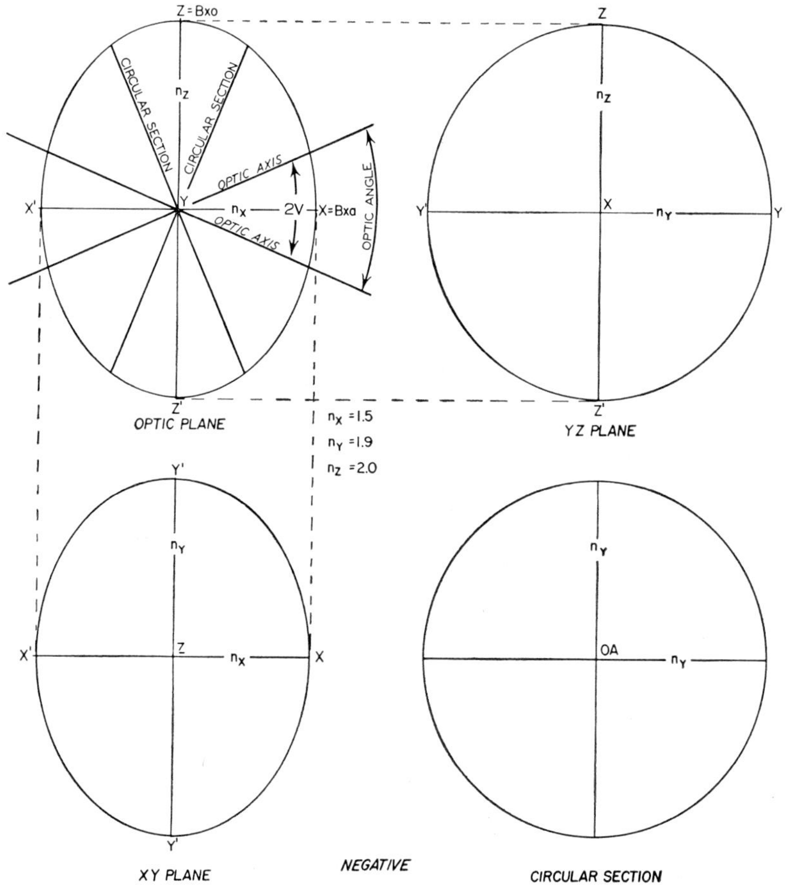

FIG. 4. Principal planes of a negative biaxial indicatrix in which $n_X = 1.5$, $n_Y = 1.9$, and $n_Z = 2.0$.

Accordingly,

$$OS{\cdot}OM = OX{\cdot}OZ = n_X{\cdot}n_Z$$

and

$$OM = \frac{n_X{\cdot}n_Z}{OS}$$

The velocity of the light vibrating in the optic plane and traveling along

the ray OS', then, is proportional to

$$\frac{1}{OM} = \frac{OS}{n_X \cdot n_Z}$$

and in Fig. 5 is indicated by the spacing of the arrows along OS'. Note that the arrows are at right angles to the wave normal in

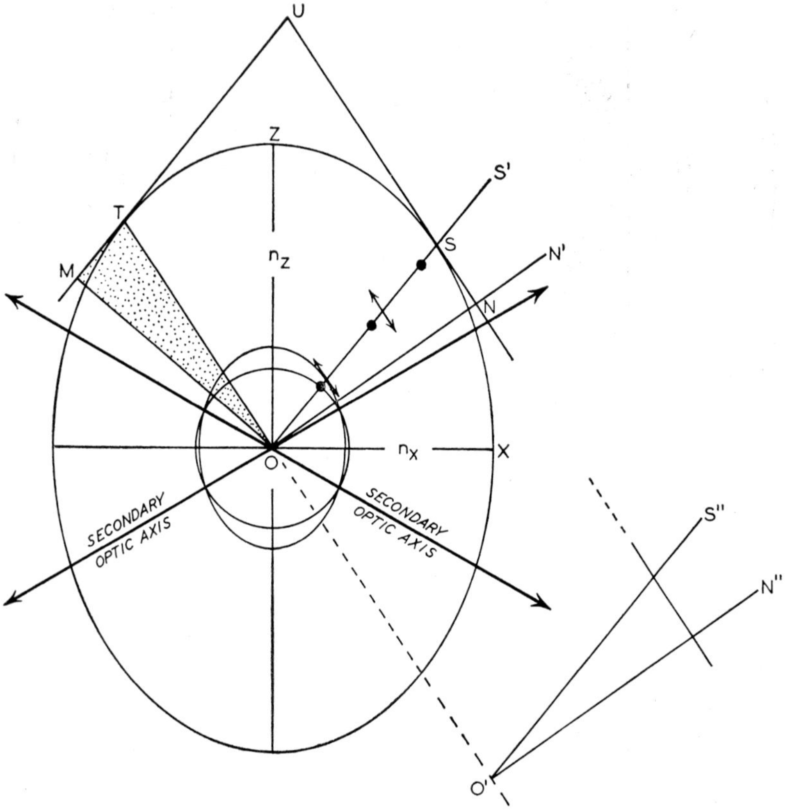

FIG. 5. Section in optic plane of a biaxial positive crystal showing relationships of waves and rays.

keeping with the electromagnetic theory of light, which states that light vibrates at right angles to the direction of transmission of the waves.

All light originating at O (Fig. 5) and moving in the directions of the radii in the XZ plane (optic plane) of the indicatrix consists of two components, one vibrating in the XZ plane and the other normal

thereto. The velocities of the components vibrating in the XZ plane are a function of the direction of vibration and propagation, and, in a given instant, light moving outward along all radii from O reaches an ellipse the semimajor and semiminor axes of which are proportional

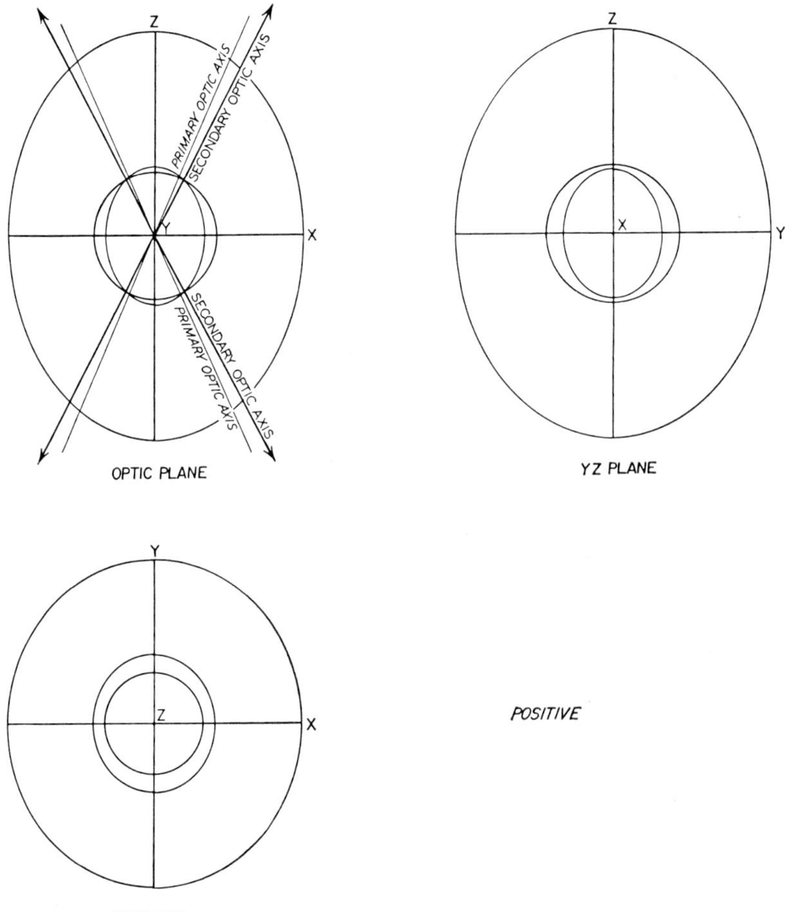

FIG. 6. Symmetry planes of a positive biaxial indicatrix showing relationship between ray velocity surfaces and indicatrix.

to $1/n_X$ and $1/n_Z$, respectively. The component vibrating normal to the XZ plane (parallel to the Y axis) has a refractive index on n_Y and a velocity proportional to $1/n_Y$. In a given instant all light vibrating normal to the XZ plane reaches a circle. In Fig. 5 the circle and smaller ellipse represent the velocities of light moving along all

RAY VELOCITY SURFACES   151

rays in the XZ plane and are sections of the ray velocity surfaces derived from the indicatrix.

In two directions, OX and OZ, the rays are parallel to the wave
normals. Accordingly, light traveling along OX consists of two com-

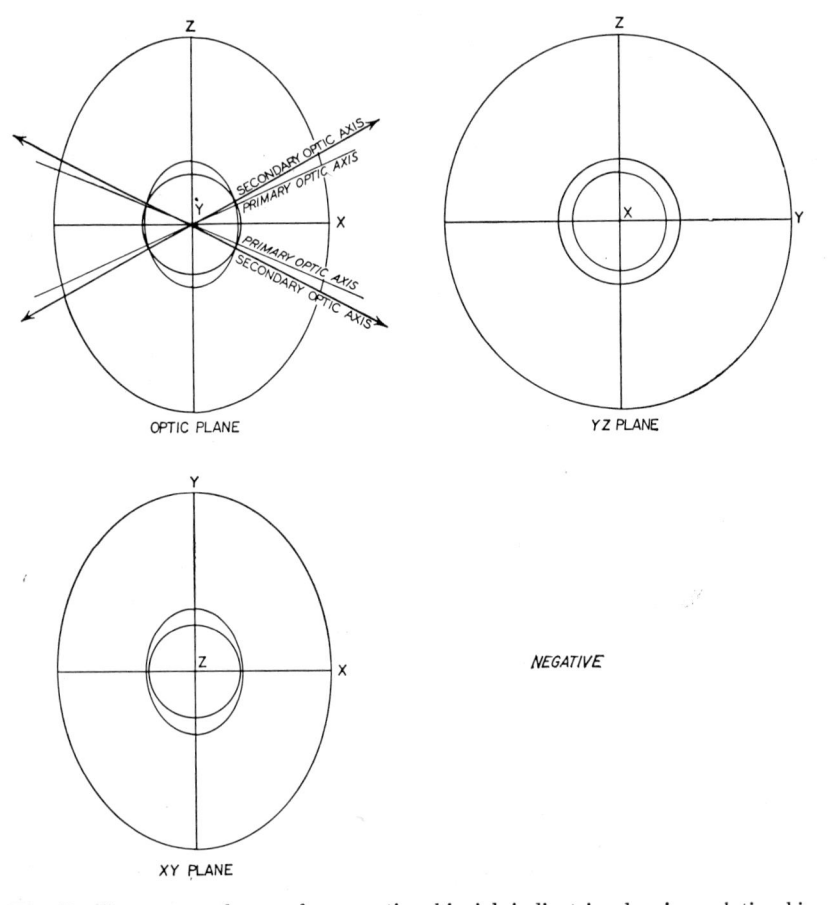

FIG. 7. Symmetry planes of a negative biaxial indicatrix showing relationship
between ray velocity surfaces and indicatrix.

ponents, one vibrating in the XZ plane and having a velocity proportional to $1/n_Z$, and the other vibrating normal to the XZ plane and
having a velocity of $1/n_Y$. The two components of the light moving
in the OZ direction have velocities proportional to $1/n_X$ and $1/n_Y$,
respectively.

The ray velocity surfaces intersect at four points in the XZ plane,
and lines drawn through these intersections give the directions of the

152   BIAXIAL CRYSTALS—THE BIAXIAL INDICATRIX

secondary optic axes. The secondary optic axes are directions of *equal ray velocity*, as opposed to the *primary optic axes*, which are normal to the circular sections of the indicatrix and are directions of *equal wave velocity*. The angle between the primary and secondary optic axes is generally less than 2 degrees.

Sections of ray velocity surfaces in the symmetry planes of positive and negative biaxial indicatrices are shown in Figs. 6 and 7.

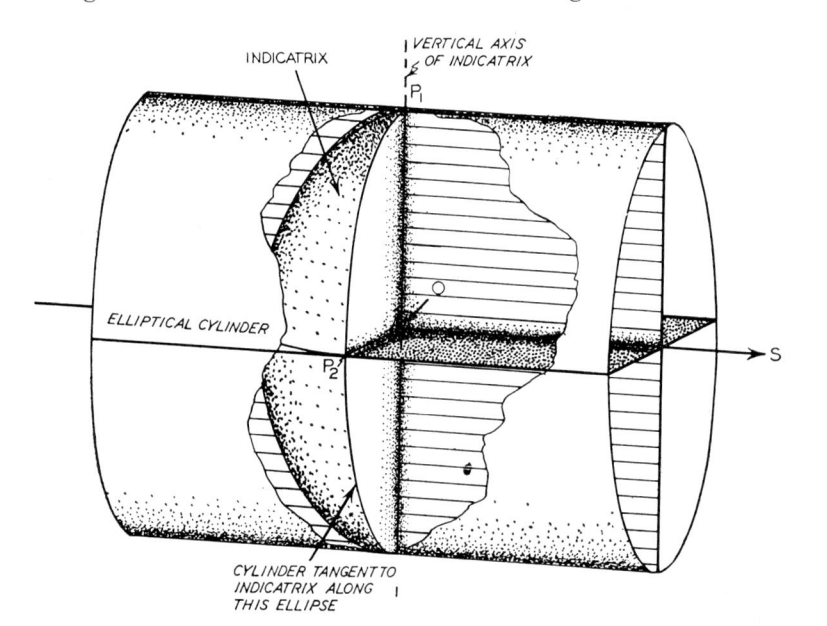

FIG. 8.   Diagram to show relationships between rays and waves traveling in a direction normal to a symmetry plane of a biaxial indicatrix.

Generalized Interrelationships of Wave Normals and Rays. The wave normal is perpendicular to the wave front and is parallel to the direction that a wave follows in passing through a crystal. A ray is the path that light follows in moving from one point to another. Both waves and the light in rays vibrate perpendicular to the wave normal. The *plane of vibration* is a plane which includes the ray and the wave normal, and the index of light propagated along a ray is the component of the wave index at right angles to the ray in the plane of vibration.

At right angles to symmetry planes of the biaxial indicatrix, rays coincide with their wave normal. In Fig. 8 suppose that OS is perpendicular to a symmetry plane of an indicatrix and is the direction of movement of light originating at O along certain rays. An

WAVE NORMALS AND RAYS

elliptical cylinder with its elements parallel to OS is drawn tangent to the indicatrix. OP_1 and OP_2 are the indices of the two components of light propagated along the ray and lie in the symmetry planes of the tangent cylinder. Moreover, the wave normal is parallel to OS, and the components of the wave moving in the direction of its wave normal also have indices equal to OP_1 and OP_2.

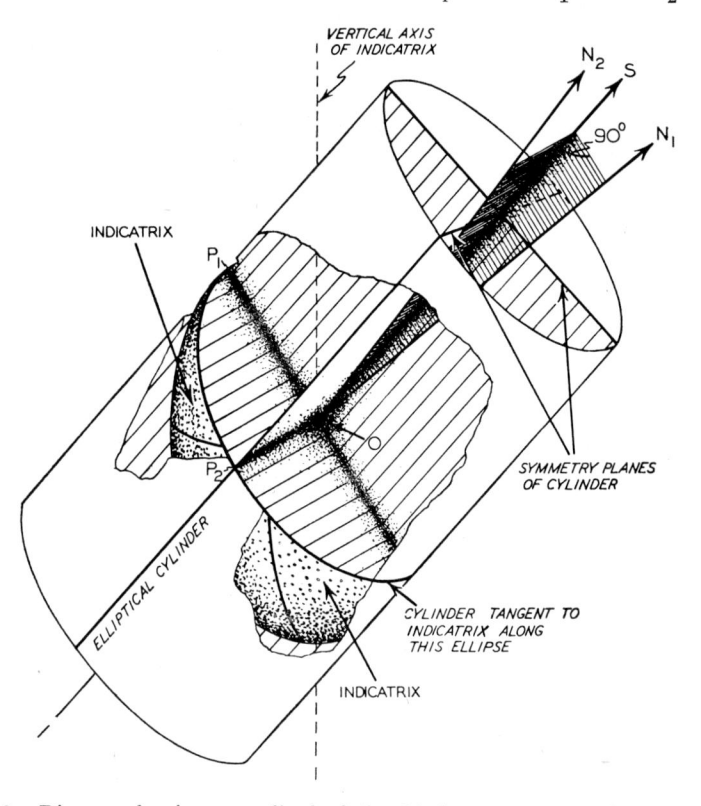

FIG. 9.   Diagram showing generalized relationship between a ray and its associated wave normals.

Figure 9 portrays a more general example. Suppose that in Fig. 9 OS is the direction of movement of light along a certain random ray, and it is desired to determine the velocities of the two components of light in the ray and to determine the positions of the associated wave normals. As in Fig. 8, a cylinder with its elements parallel to OS is drawn tangent to the indicatrix. This cylinder has an elliptical cross section for all positions except those in which the points of tangency with the indicatrix coincide with a circular section.

The semimajor and semiminor axes of a section of the cylinder at

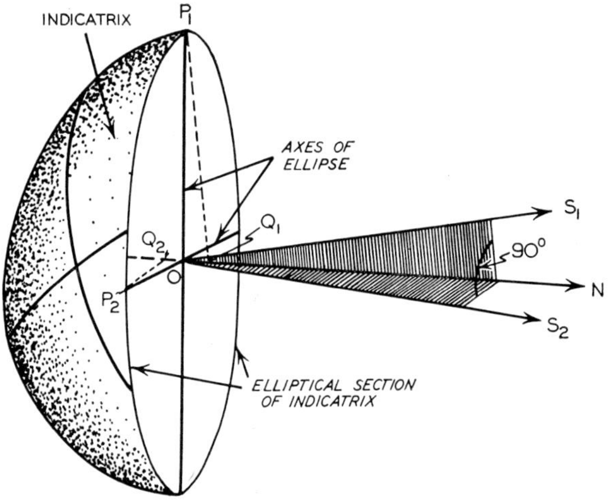

FIG. 10. Diagram showing relationship between a wave normal and its associated rays.

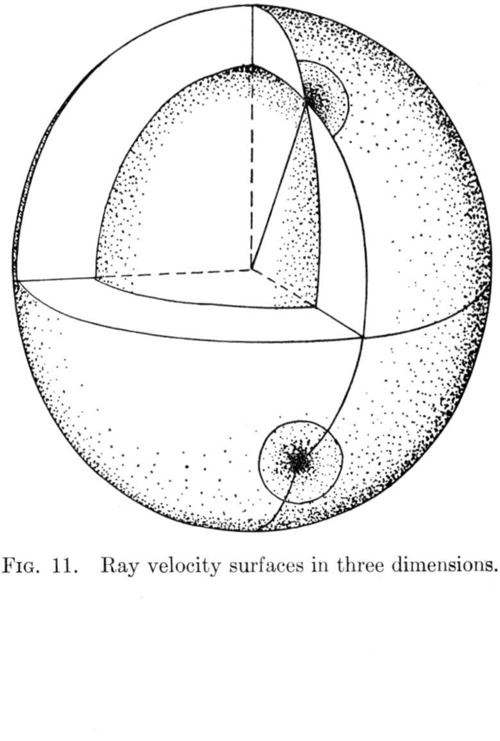

FIG. 11. Ray velocity surfaces in three dimensions.

right angles to OS give the refractive indices of the two components of light traveling along OS. The velocities of these components are inversely proportional to their refractive indices.

The ellipse determined by the points of tangency of the cylinder and the indicatrix is a section of the indicatrix and is not parallel to

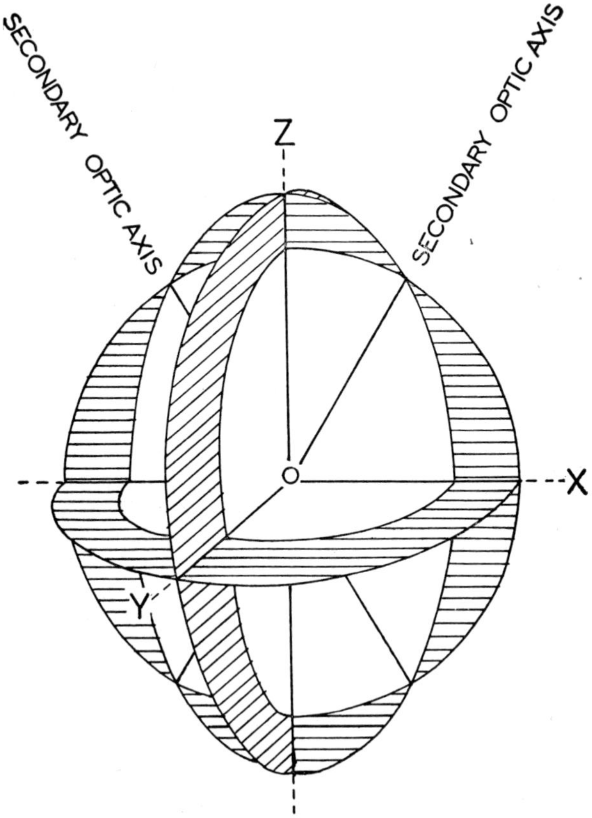

FIG. 12. Symmetry planes of ray velocity surfaces.

the cross section of the tangent cylinder. OP_1 and OP_2 are dimensions of the indicatrix determined by the intersection of the symmetry planes of the cylinder and the elliptical plane through the indicatrix, and they give the refractive indices of the waves corresponding to the two components of light following the ray OS. Inasmuch as waves vibrate perpendicular to their normals, directions ON_1 and ON_2, in the symmetry planes of the cylinder, give the directions of the wave normals of the waves corresponding to the two components of light in ray OS. The waves have velocities of $1/OP_1$ and $1/OP_2$, respectively.

Now, suppose that the direction of the wave normal is known, and it is desired to locate the directions of the rays corresponding to the two components of the wave. In Fig. 10, ON is the wave normal perpendicular to a random elliptical section of an indicatrix. The semimajor and semiminor axes of this section, OP_1 and OP_2, give the refractive indices of the two components of the wave moving along ON. The velocities of the components of the wave are proportional to $1/OP_1$ and $1/OP_2$, respectively.

To obtain the refractive indices of the light in the rays, normals are dropped to ON from P_1 and P_2. P_1Q_1 and P_2Q_2 are the ray indices of refraction corresponding to the light following the two rays. OS_1 and OS_2, drawn perpendicular to P_1Q_1 and P_2Q_2 in planes including OP_1 and OP_2 and the directions of the wave normal ON, are the directions of the rays. The velocities of the light along these rays are $1/P_1Q_1$ and $1/P_2Q_2$, respectively.

When the velocity for light moving along all rays in a crystal has been determined, it becomes possible to construct three-dimensional ray velocity surfaces such as those shown in Fig. 11. The ray velocity surfaces intersect at four points—at the bottoms of four "dimples" in the surface of the outer shell. The directions obtained by drawing lines through the bottoms of the dimples and the center are directions of the secondary optic axes. The dimples are outlined by circles which represent the line of contact of the outer shell and a plane tangent to the shell.

Figure 12 portrays the sections of the ray velocity surfaces in the planes of symmetry.

Optic Orientation of Orthorhombic Crystals. The three planes of symmetry of the indicatrix demand that certain conditions regarding its orientation be fulfilled in biaxial crystals. That is, the position of the indicatrix with reference to the symmetry elements of a crystal cannot be such as to lower the symmetry of that crystal. Optical properties as related to crystallographic directions must show the same geometrical distribution or variations as other physical properties, such as hardness, luster, or cleavage.

Orthorhombic crystals have three mutually perpendicular unequal axes and, in the dipyramidal class, three planes of symmetry. This symmetry requires the indicatrix to be oriented so that its three mutually perpendicular axes, the X, Y, and Z axes, coincide with the a, b, and c axes of the crystal. Any deviation from this principle would result in lowering the symmetry of the crystal, thus placing it in the monoclinic or triclinic system. The parallel orientation of indicatrix axes and crystallographic axes results in coincidence

of the planes of symmetry. Moreover, the optic plane must lie in one of the three crystallographic planes of symmetry.

The following orientations are possible:

1. Optic plane parallel to (010) and

$$\begin{aligned} X &= a \\ Y &= b \\ Z &= c \end{aligned} \quad \text{or} \quad \begin{aligned} X &= c \\ Y &= b \\ Z &= a \end{aligned}$$

2. Optic plane parallel to (100) and

$$\begin{aligned} X &= b \\ Y &= a \\ Z &= c \end{aligned} \quad \text{or} \quad \begin{aligned} X &= c \\ Y &= a \\ Z &= b \end{aligned}$$

3. Optic plane parallel to (001) and

$$\begin{aligned} X &= a \\ Y &= c \\ Z &= b \end{aligned} \quad \text{or} \quad \begin{aligned} X &= b \\ Y &= c \\ Z &= a \end{aligned}$$

Figure 13 illustrates an orthorhombic crystal for which the orientation and certain other optical data for a particular wave length may be expressed as follows:

	n	*Orientation*	*Bisectrices*	
X	n_X	c	*Bxa*	*Negative*
Y	n_Y	a	—	$2V = 18°$
Z	n_Z	b	*Bxo*	

Optic Orientation of Monoclinic Crystals. Monoclinic crystals have three unequal axes. The a and c axes intersect at acute and obtuse angles and lie, in the prismatic class, in a plane of symmetry. The b axis is normal to the symmetry plane. The symmetry of a monoclinic crystal requires the following possibilities with respect to the orientation of the indicatrix for a single wave length:

1. The X axis is parallel to the b crystallographic axis. The optic plane is parallel to the b axis but may occupy any position with respect to the a and c axes, but only one position for monochromatic light in a given crystal species.

2. The Y axis parallels the b crystallographic axis. The optic plane lies in the (010) plane, and the X and Z axes, mutually perpendicular, may lie in any position with respect to the a and c axes, but in a definite position for each crystalline substance.

3. The Z axis parallels the b crystallographic axis, and the optic plane is parallel to the b axis and may occupy any position with respect to the a and c axis, but only one position for a given crystal.

Figure 14 shows one possibility for a monoclinic crystal in monochromatic light. The Y axis of the indicatrix is parallel to the b

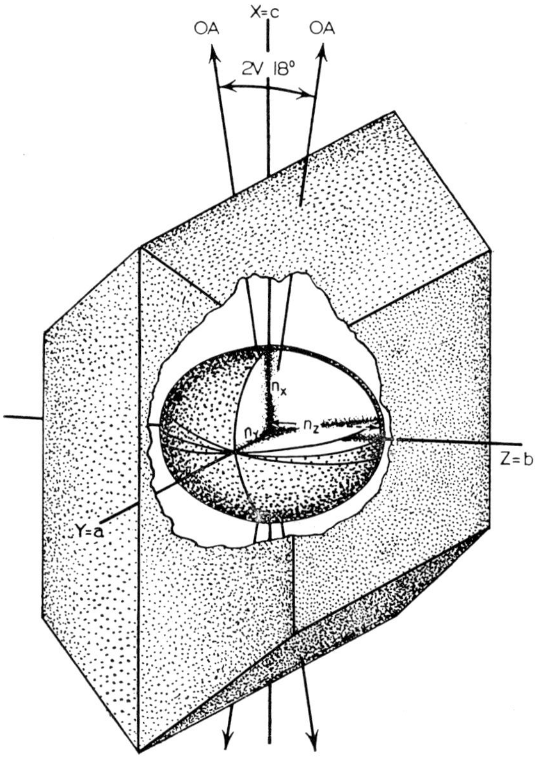

FIG. 13. Negative orthorhombic crystal in which $X = c$, $Y = a$, and $Z = b$. Optic plane is parallel to (100).

axis of the crystal. The optic orientation and certain other data for monochromatic light may be tabulated as follows:

	n	*Orientation*	
X	n_X		*Negative*
Y	n_Y	b	$2V = 80°$
Z	n_Z	$\wedge c = 16°$	

Optic Orientation of Triclinic Crystals. Triclinic crystals have no planes of symmetry. Accordingly, the indicatrix may occupy any position with respect to the crystallographic axes, but the position is fixed for a given crystal and a particular wave length of light.

The orientation of the indicatrix in triclinic crystals is seen to best advantage in stereographic projections. Measurements commonly are

made with a universal stage, and the results are plotted so as to show the angular relationships between prominent crystal faces and the important directions of the indicatrix. By means of a stereographic net it is possible to rotate a projection into any desired position.

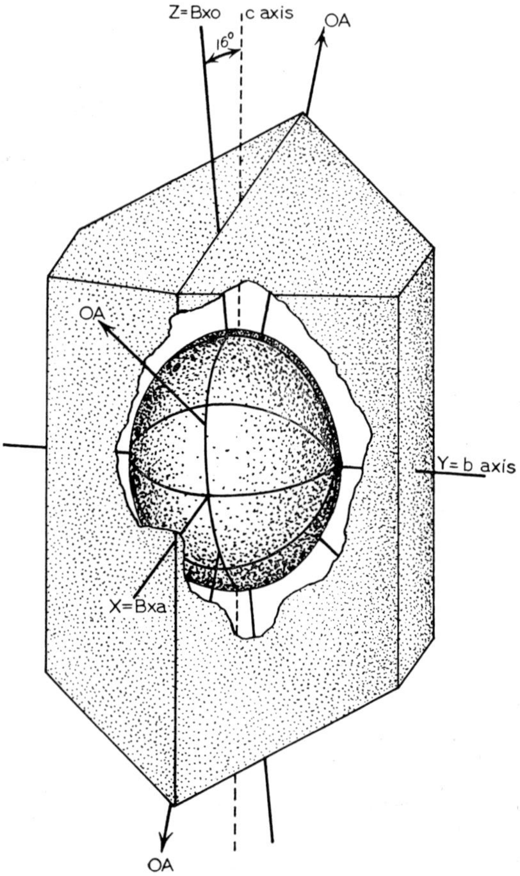

Fig. 14. Monoclinic crystal showing the orientation of the indicatrix for a particular wave length of light. Optic plane is parallel to (010). $Z \wedge c = 16°$.

Crystallographers use ϕ and ρ to indicate the angular positions of linear directions or the normals to crystal faces. ϕ is the angle of azimuth measured from the zero meridian [commonly the pole of the (010) face of a crystal] and is positive in clockwise direction from 0 to 180 degrees. In the counterclockwise direction from the zero meridian, ϕ is negative. ρ is the angular inclination of a direction or a normal to a face with respect to the north-south axis of the spherical

projection and is plotted as the angular distance of a pole from the center of the projection.

The crystallographic study of axinite by Peacock[1] provides an excellent example of the interrelationships of optical and other crystal-

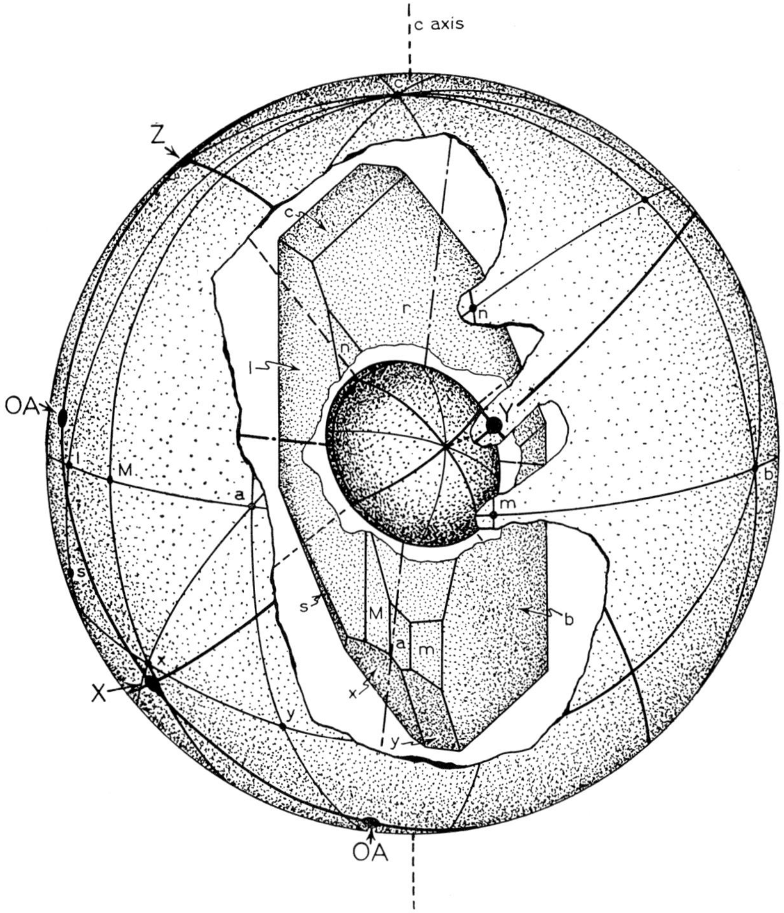

FIG. 15. Spherical projection of crystallographic data for axinite. Both the sphere and the crystal of axinite inside it have been cut away to show in three dimensions the interrelationships of the spherical projection, the axinite crystal, and the indicatrix for sodium light. Crystallographic data from Peacock: *Am. Mineral.*, 1937.

lographic constants. The accompanying table is based on Peacock's data. Figure 15 illustrates the relationships expressed in the table in a three-dimensional drawing.

[1] M. A. Peacock, "On the crystallography of axinite . . . ," *Am. Mineralogist,* Vol. 22, pp. 588–624, 1937.

CRYSTALLOGRAPHIC CONSTANTS FOR AXINITE

(Selected from more complete data by Peacock)

Triclinic; pinacoidal

Axial angles: $\alpha = 91°\ 51\frac{1}{2}'$; $\beta = 98°\ 04'$; $\gamma = 77°\ 14'$

Forms		ϕ	ρ	*Forms*		ϕ	ρ
c	001	89° 23	8° 04′	*r*	011	8 03	45 21
b	010	0 00	90 00	*y*	$\bar{1}01$	−75 45	49 13
a	100	102 38	90 00	*n*	111	62 51	57 41
m	110	60 28	90 00	*x*	$\bar{1}11$	−41 09	59 39
M	$1\bar{1}0$	135 25	90 00	*s*	$\bar{1}21$	−26 10	68 34
l	$1\bar{2}0$	151 01	90 00				

Optical Elements

	ϕ	ρ	$n_{(Na)}$		
X	−42°	56°	1.683	±0.002	*Negative* $2V = 81°$
Y	59	75	1.688		
Z	168	39	1.692		

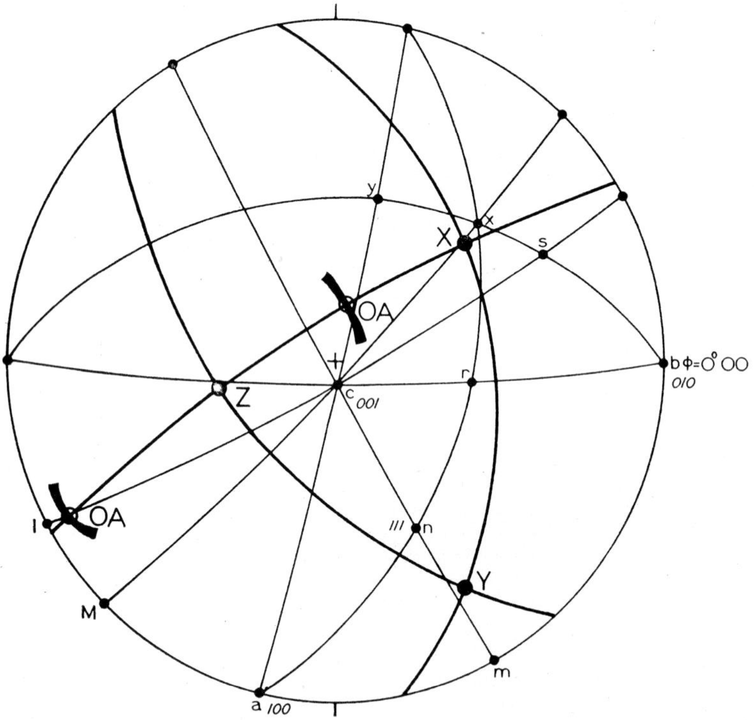

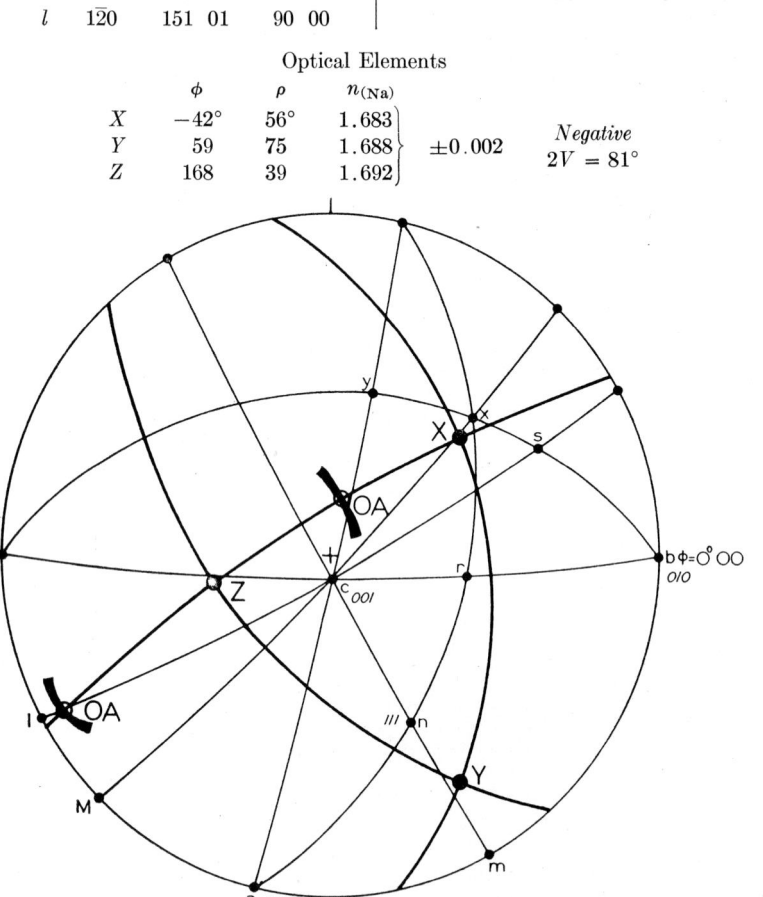

FIG. 16. Stereographic projection of optical and other crystallographic data for axinite. Plane of projection is the equatorial plane of the spherical projection shown in Fig. 15.

In Fig. 15 windows have been cut away to show the orientation of the indicatrix for sodium light in an axinite crystal, the center of

which is coincident with the center of a hollow sphere enveloping the crystal. Projection of the optical directions and the normals to the crystal faces to the surface of the sphere gives a spherical projection of the poles of the directions and crystal faces. The great circles passing through the poles are zonal lines on the surface of the sphere.

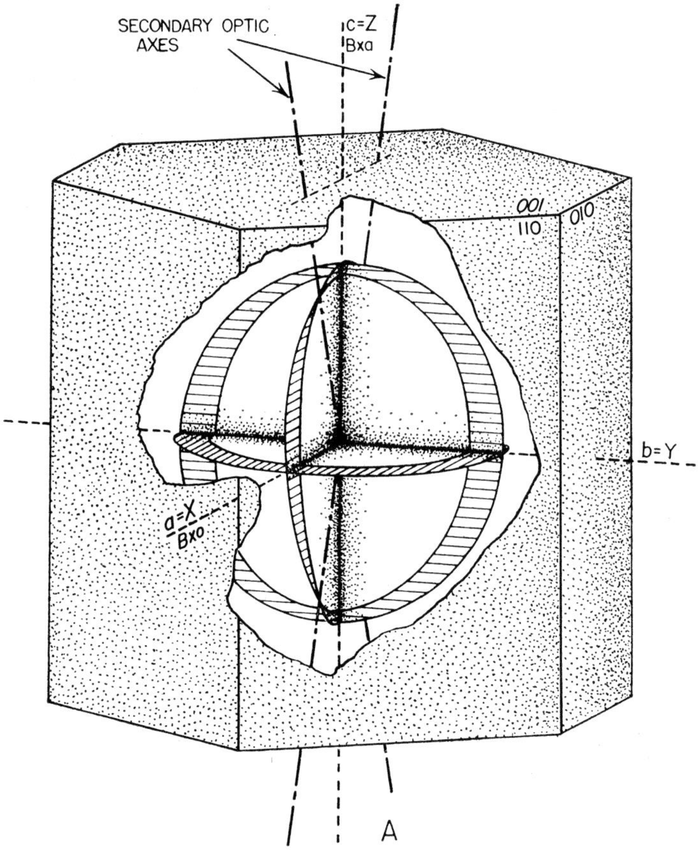

FIG. 17. Huygenian constructions in a positive orthorhombic crystal.

$X = a;\ Y = b;\ Z = c.$

A. Crystal cut away to show sections in symmetry planes of ray velocity surfaces.
B. Light falling with inclined incidence on (010). Plane of incidence parallel to (001).
C. Light falling with inclined incidence on (001). Plane of incidence parallel to (010).
D. Light falling with perpendicular incidence on (110). Plane of incidence parallel to (001).

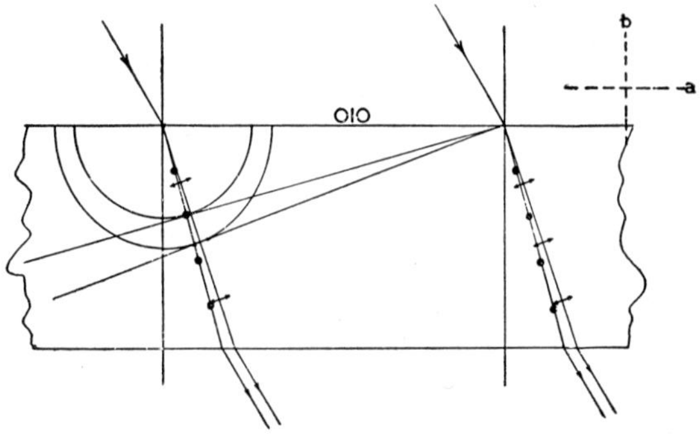

B

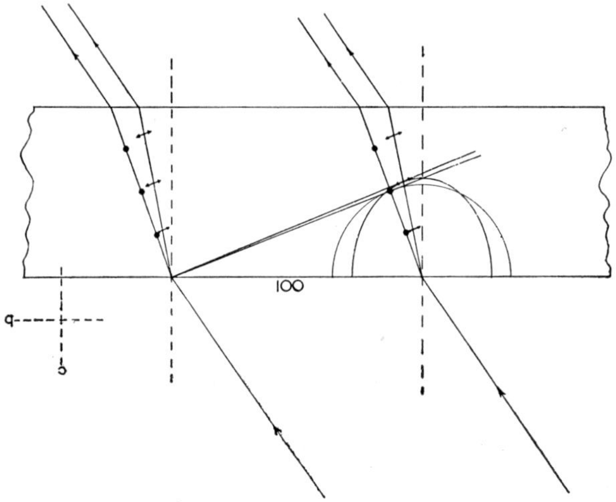

C

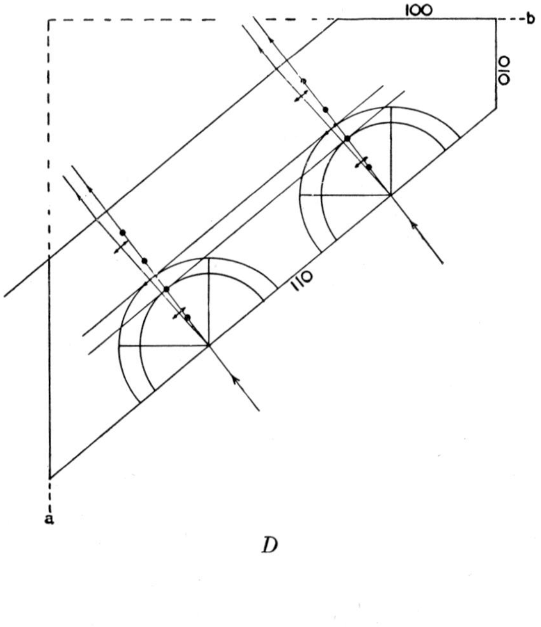

D

Figure 16 is a stereographic projection of the same data into the equatorial plane of the spherical projection in Fig. 15. The brushes passing through the points of emergence of the optic axes (OA) suggest that the X axis is the acute bisectrix. The zones shown in Fig. 16 correspond to those shown on the surface of the spherical projection in Fig. 15.

Huygenian Constructions in Biaxial Crystals. Figure 17 shows Huygenian constructions in several sections of an orthorhombic crystal. Figure 17*A* is a clinographic projection of the crystal cut away to show the symmetry planes of the ray velocity surfaces corresponding to the designated optical orientation of the crystal. The sections in Figs. 17*B*, *C*, and *D* are parallel to planes of symmetry of the crystal. Any section other than one including two axes of the indicatrix would require a three-dimensional drawing because the rays for refracted light do not lie in the same plane, although the planes including the vibration directions are mutually perpendicular. In this particular crystal, the orientation is $X = a$; $Y = b$; $Z = c$. The crystal is optically positive.

In the drawings the relative velocities of the components of light moving along the rays are indicated by the spacing of the dots and arrows. The arrows indicate the direction of vibration of components whose ray velocity surfaces in the plane of the drawing are ellipses. The arrows are not perpendicular to the direction of propagation in the crystal, but, rather, parallel to the tangent to the ray velocity surface at the point of intersection with the ray. The components vibrating normal to the drawing are indicated by dots. These components have circular ray velocity surfaces in the sections of the drawings.

Determination of Refractive Indices and Optic Orientation of Biaxial Crystals. The three principal indices of refraction of biaxial crystals are most conveniently determined by the immersion method; n_X and n_Z may be measured in fragments showing a maximum interference color between crossed polarizing prisms. Such fragments lie so that the Y axis of the indicatrix is parallel to the axis of the microscope. The fragment is rotated to extinction, the analyzing prism removed, and the index is compared with that of the immersion medium. At the extinction position, the vibration direction of the lower prism is parallel to either X or Z. After one index is obtained, the stage is rotated 90 degrees to obtain the other.

To measure n_Y, the intermediate index, a fragment is used which remains black or dark gray during a complete rotation of the microscope stage. This fragment lies so that one of its optic axes is parallel

ABSORPTION OF LIGHT BY BIAXIAL CRYSTALS 165

to the axis of the microscope. Light traveling along an optic axis has an index equal to n_Y.

Cleavages may prevent crystal fragments from lying in a desired position, and it may become necessary to roll the fragments or to support them with ground cover glass.

Interference figures are helpful in determining the orientation of the indicatrix and assist materially in index determinations. The relationship of the indicatrix to interference figures is discussed in subsequent chapters.

To determine optic orientation it is necessary to ascertain the position of the indicatrix with reference to prominent crystal directions, usually the crystal axes. In orthorhombic crystals the X, Y, and Z axes of the indicatrix are parallel to the crystallographic axes, but in monoclinic crystals only one axis is parallel to a crystallographic axis, and in triclinic crystals none of the axes of the indicatrix coincide with the crystallographic axes.

For example, suppose that in a monoclinic crystal with prismatic (110) cleavage the extinction angle of a fragment showing a maximum interference color is 15 degrees, measured from the c axis as determined by intersecting cleavage surfaces. In the extinction position the grain is lying so that X or Z is parallel to the plane of polarization of the lower polarizing prism. Measurement of the refractive index tells whether it is X or Z, or, as an alternative, the grain is rotated to the 45-degree position from the extinction position and the relative velocities of the components are determined by means of an accessory plate. If the Z direction makes an angle of 15 degrees with the direction of the intersection of the cleavage, the relationship is expressed by the equation $Z \wedge c = 15$ degrees. When X and Z lie in the symmetry plane of the crystal, Y of necessity lies parallel to the b crystallographic axis.

Interference Colors. Interference colors in biaxial crystals, as in uniaxial crystals, depend on the thickness, orientation, and birefringence of crystal plates or fragments. Abnormal interference colors result from differential absorption of light or may be an expression of the fact that the crystal is optically abnormal in certain respects. Certain rare substances may be uniaxial for one color of the spectrum and biaxial for the other colors, or the substance may be isotropic for one color and biaxial for others.

Absorption of Light by Biaxial Crystals. Differential absorption of light by crystals results in *pleochroism*. Some biaxial crystals, when viewed in plane-polarized light, change color during rotation of the microscope stage. Ordinarily, the pleochroism of absorption is determined in the directions of the X, Y, and Z axes of the indicatrix.

166   BIAXIAL CRYSTALS—THE BIAXIAL INDICATRIX

Thus the absorption formula of a certain crystal might be

X = weak

Y = strong

Z = very strong

or the pleochroism might be expressed in the following terms:

X = pale green

Y = dark green

Z = dark brown

It might be possible to express color variation in terms of a triaxial ellipsoid which would indicate the positions of the maximum, minimum, and intermediate absorption directions. In orthorhombic crystals the axes of the absorption ellipsoid would parallel crystallographic axes. In monoclinic crystals one axis would parallel the b crystallographic axis, and the other directions would lie in the ac plane. However, the absorption ellipsoid might not coincide with the indicatrix, thus making the determination of pleochroic colors in the directions of the indicatrix a purely artificial procedure. In triclinic crystals, the absorption ellipsoid again would not necessarily be in a parallel position with the indicatrix.

Interior and Exterior Conical Refraction. Interior and exterior conical refraction are of secondary importance in the practical application of crystal optics, but are of considerable theoretical interest and serve to demonstrate certain relations between the indicatrix and ray velocity surfaces. In the development of the subsequent theory it should be remembered that the primary optic axis is a *direction of equal wave velocity,* and the secondary optic axis is a *direction of equal ray velocity.*

Interior conical refraction can be demonstrated by the experiment suggested in the three-dimensional drawing in Fig. 18A. A plate of a biaxial substance is cut exactly at right angles to the primary optic axis (parallel to a circular section of the indicatrix), and monochromatic light is allowed to fall with perpendicular incidence on a pinhole in a metal sheet covering the base of the plate All of the light does not travel straight through the plate as might be expected, but is resolved into a hollow cone with its apex at the pinhole. Upon leaving the crystal plate, the light in the cone is again refracted so as to form a hollow cylinder, the diameter of which is equal to the diameter of the base of the interior cone. The presence of the hollow cylinder is determined by holding a flat surface in the path of the cylinder and observing

INTERIOR AND EXTERIOR CONICAL REFRACTION   167

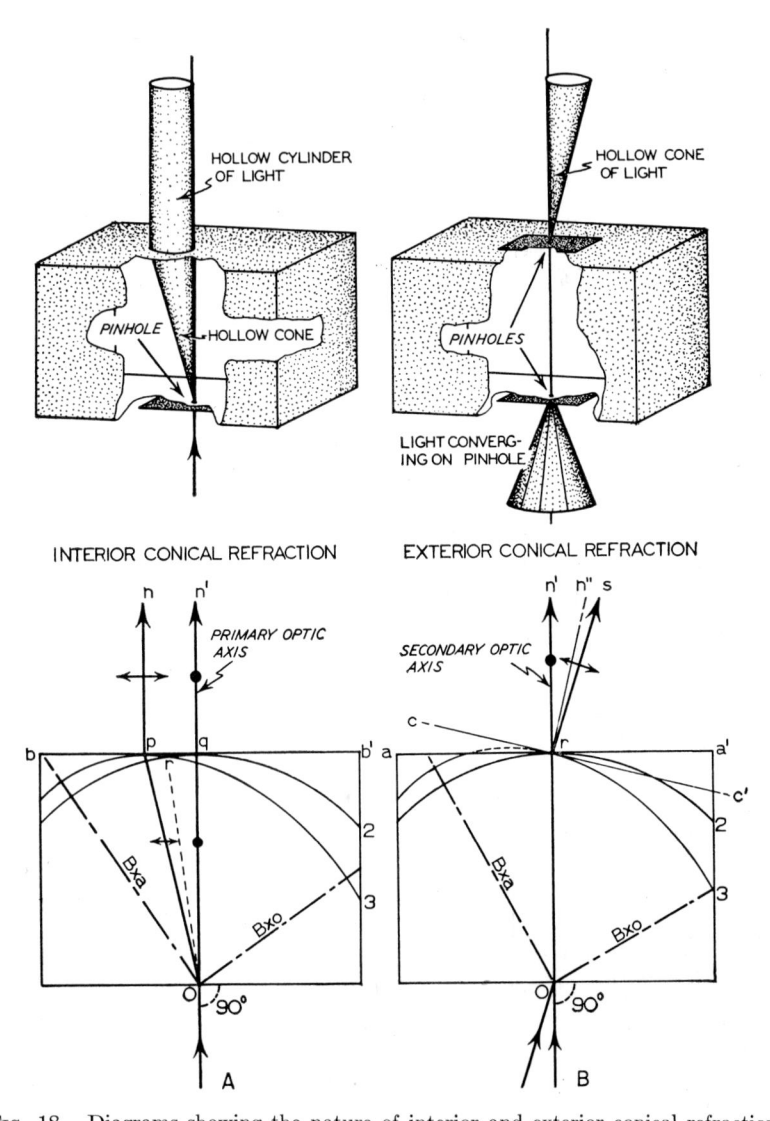

FIG. 18.   Diagrams showing the nature of interior and exterior conical refraction.

A.   Interior conical refraction in three dimensions and cross section.

B.   Exterior conical refraction in three dimensions and cross section.

a ring of light, the diameter of which does not change as the surface is raised or lowered.

The explanation of interior conical refraction is suggested in the line drawing in Fig. 18*A*. The plane of the drawing is the optic plane. Curves 2 and 3 are sections of ray velocity surfaces; *Or* is the direction of the secondary optic axis; *Oq* is the direction of the primary optic axis and is perpendicular to *bb′*, the wave front tangent to the ray velocity surfaces at *p* and *q*. In three dimensions the wave front

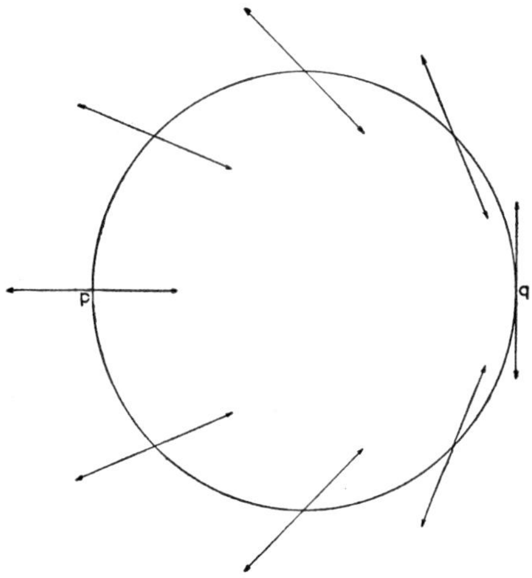

FIG. 19. Traces of vibration planes of light in interior conical refraction.

is tangent to the outer shell of the ray velocity surfaces along a circle marking the lip of the dimple at the point of emergence of a secondary optic axis (see Fig. 11).

A wave of light perpendicularly incident at *O* travels through the crystal in a direction parallel to *Oq*, the wave normal. Light in rays associated with the wave travels along either *Op* or *Oq*, depending upon the vibration direction. Despite the difference in the paths and velocities of light moving along the rays, all of the light reaches the wave front *bb′* at the same instant. In three dimensions this light constitutes a hollow cone. The traces of the planes of vibration of the light in the rays in the cone are shown in plan in Fig. 19. Light leaving the crystal plate is refracted so as to be parallel to the incident light and thus constitutes a hollow cylinder above the plate.

Exterior conical refraction is demonstrated by the experiment shown in the three-dimensional drawing in Fig. 18*B*. A crystal plate is cut exactly at right angles to the secondary optic axis of a biaxial crystal.

Pinholes on the upper and lower surfaces of the plate are aligned so that only the light passing through the plate in the direction of the secondary optic axis can emerge. A cone of monochromatic light is caused to converge on the pinhole on the lower surface. The light in the cone will be refracted in all directions in the plate, but only the light that is refracted so as to pass along the secondary optic axis emerges from the plate through the upper pinhole.

Under the above conditions the emergent light forms a hollow cone *outside* the crystal plate. The apex of the cone is at the upper pinhole. The presence of the cone can be demonstrated by placing a flat surface parallel to the plate in the path of the emergent light, so as to produce a ring at right angles to the axis of the cone. As the flat surface is raised or lowered, the diameter of the ring changes, thus demonstrating that the light actually comprises a cone.

An explanation of exterior conical refraction is offered in the line diagram in Fig. 18*B*, a section in the optic plane. Light converging at O and refracted so as to travel along the secondary optic axis emerges at the pinhole at r. Or is the direction of the secondary optic axis, and aa' and cc' are wave fronts in the crystal and are tangent to the ray velocity surfaces 2 and 3 at r. Curve 2 represents the velocities of light vibrating perpendicular to the plane of the drawing, and curve 3 represents the velocities of light vibrating in the plane of the drawing; rn' is the wave normal in the crystal for the wave front aa', and rn'' is the wave normal for the wave front cc'.

Light vibrating in a ray perpendicular to the plane of the drawing travels in the same direction as its wave normal On' and emerges from the plate without refraction. Light vibrating in the plane of the drawing is refracted upon leaving the plate and makes a small angle with rn''. If light vibrating in the plane of the drawing had a ray index of unity, its path on emergence would be parallel to rn'', but, because this is not so, it should be expected that the wave normal in the crystal and the wave normal of light in the emergent ray are not coincident. In effect rn' and rs are the wave normals of the emergent light in the plane of the drawing.

All light emerging at r constitutes a hollow cone, the so-called *cone of external refraction*. Inasmuch as the laws of refraction are operative, the light in the exterior cone corresponds to the light in the incident cone having angles of incidence equal to the angles of refraction of the emergent light.

The above explanations hinge on the fact that the directions of the primary and secondary optic axes do not exactly coincide in biaxial crystals. The angle between the two types of optic axes is generally

very small—less than two degrees. Moreover, the apical angles of the cones of interior and exterior conical refraction are of the same order of magnitude. One of the most noticeable effects of refraction of either type is seen in crystal grains or sections cut at right angles to an optic axis. Such grains are not completely dark in all positions between crossed polarizing prisms, a fact which demonstrates that a slight path difference is present among the various components of light moving through a crystal in the general direction of an optic axis.

From a theoretical point of view, interior and exterior conical refraction demonstrate the validity of the ray and wave concepts of light movement through crystals.

Light Surfaces Related to the Indicatrix. The indicatrix and the ray velocity surfaces provide the most useful geometric bases for the optical study of crystals without rotatory power. But it is possible to represent the geometry of light propagation in crystals by one or all of six surfaces, all rigidly interrelated mathematically. Although modern investigators use the indicatrix and ray velocity surface concepts almost exclusively, older writers have used one or all of four other concepts. An understanding of these is of value in the development of a comprehensive background of crystal optics. The following is a description of each of the six related surfaces.

1. The *indicatrix* is a triaxial, three-dimensional ellipsoidal envelope, the three semiaxes of which correspond to the indices of refraction of waves of monochromatic light in a biaxial crystal in their *direction of vibration.* This is emphasized in foregoing discussions.

The general equations for the triaxial ellipsoid are as follows:

In the optic or XZ plane

$$\frac{X^2}{(n_X)^2} + \frac{Z^2}{(n_Z)^2} = 1$$

or in three dimensions

$$\frac{X^2}{(n_X)^2} + \frac{Y^2}{(n_Y)^2} + \frac{Z^2}{(n_Z)^2} = 1$$

2. The *ray velocity surfaces* indicate the velocity of monochromatic light in its direction of propagation along rays. The ray velocity surfaces are derived from the indicatrix, as explained on page 146. The equations in the three symmetry planes of the ray velocity surfaces are:

a. In the XZ section for the ellipse

$$\frac{X^2}{(1/n_Z)^2} + \frac{Z^2}{(1/n_X)^2} = 1$$

and for the circle

$$X^2 + Y^2 = \left(\frac{1}{n_Y}\right)^2$$

b. In the XY section for the ellipse

$$\frac{X^2}{(1/n_Y)^2} + \frac{Y^2}{(1/n_X)^2} = 1$$

and for the circle

$$X^2 + Y^2 = \left(\frac{1}{n_Z}\right)^2$$

c. In the YZ section for the ellipse

$$\frac{Y^2}{(1/n_Z)^2} + \frac{Z^2}{(1/n_Y)^2} = 1$$

d. For the double-shelled surface in three dimensions

$$\frac{X^2}{X^2 + Y^2 + Z^2 - (1/n_X)^2} + \frac{Y^2}{X^2 + Y^2 + Z^2 - (1/n_Y)^2} + \frac{Z^2}{X^2 + Y^2 + Z^2 - (1/n_Z)^2} = 1$$

3. The two-shelled *wave velocity surface* indicates the velocity of light waves in the *direction of propagation*, that is, in the directions of the wave normals. This surface is derived from the ray velocity surface.

The equation for the wave velocity surface in the XZ plane is

$$(X^2 + Z^2)^2 = \frac{X^2}{(n_Z)^2} + \frac{Z^2}{(n_X)^2}$$

In three dimensions

$$\frac{X^2}{X^2 + Y^2 + Z^2 - (1/n_X)^2} + \frac{Y^2}{X^2 + Y^2 + Z^2 - (1/n_Y)^2} + \frac{Z^2}{X^2 + Y^2 + Z^2 - (1/n_Z)^2} = 0$$

4. The *Fresnel ellipsoid* or the *vibration velocity surface* is a single surface in the form of a triaxial ellipsoid. The ellipsoid indicates the velocity of light along rays in its *direction of vibration*. The three principal axes are equal respectively to $1/n_X$, $1/n_Y$, and $1/n_Z$.

In three dimensions the equation is

$$\frac{X^2}{(1/n_X)^2} + \frac{Y^2}{(1/n_Y)^2} + \frac{Z^2}{(1/n_Z)^2} = 1$$

5. The *ovaloid* is obtained by construction from the Fresnel ellipsoid in the same way that the wave velocity surface is derived from the ray velocity surface. The ovaloid is a single surface enveloping the Fresnel

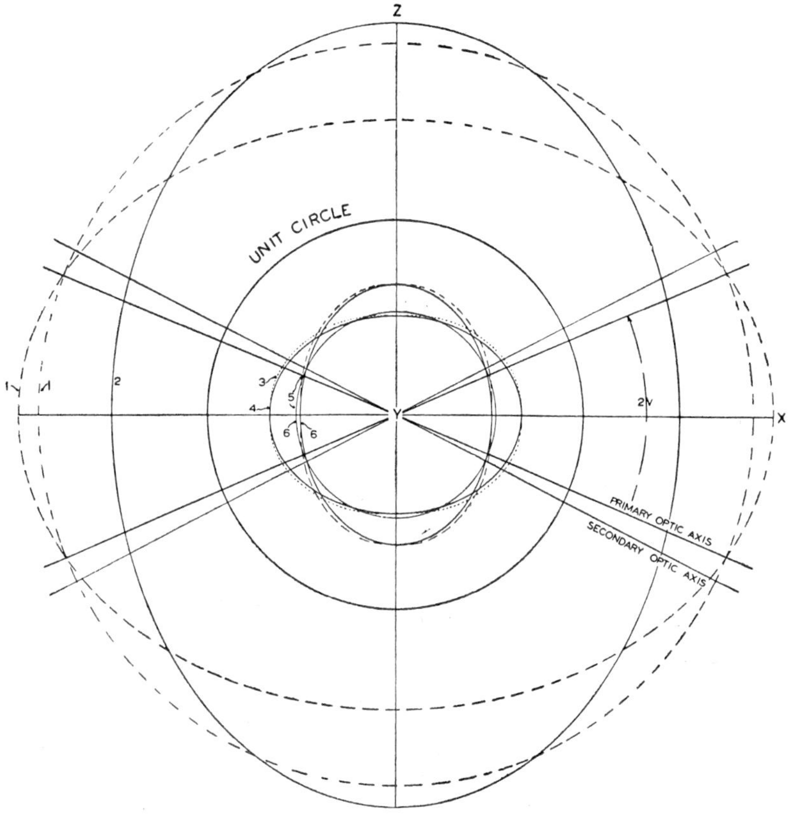

FIG. 20. Six light surfaces in biaxial negative crystal.

1. Index surfaces.
2. Indicatrix.
3. Ovaloid.
4. Fresnel surface.
5. Wave velocity surfaces.
6. Ray velocity surfaces.

ellipsoid; it represents the velocity of waves in *their direction of vibration*. Its equation is

$$(X^2 + Y^2 + Z^2)^2 - \left[\left(\frac{1}{n_X}\right)^2 X^2 + \left(\frac{1}{n_Y}\right)^2 Y^2 + \left(\frac{1}{n_Z}\right)^2 Z^2\right] = 0$$

6. The *index surface* is double-shelled and represents the indices of refraction of waves in their *directions of propagation*. In the XZ section the index surface consists of a circle and an ellipse. The circle indicates

the index of the n_Y wave in its direction of propagation. The ellipse has the same dimensions as the XZ section of the indicatrix but is rotated 90 degrees.

Its equation is

$$\frac{X^2}{X^2 + Y^2 + Z^2 - (n_X)^2} + \frac{Y^2}{X^2 + Y^2 + Z^2 - (n_Y)^2} + \frac{Z^2}{X^2 + Y^2 + Z^2 - (n_Z)^2} = 1$$

Figure 20 is a drawing showing the six types of surfaces and their interrelationship in the XZ plane. In the positive crystal illustrated the indices of refraction are arbitrarily assigned values as follows: n_X, 1.5; n_Y, 1.6; n_Z, 2.0.

CHAPTER XIV

BIAXIAL CRYSTALS IN CONVERGENT POLARIZED LIGHT

Introduction. The action of the conoscope on biaxial crystals will be more clearly understood if the manner of passage of light through the petrographic microscope (orthoscope) is reviewed briefly. In the absence of the accessory substage lens of short focal length, the plane-polarized light reaching the crystal plate on the microscope stage follows rays which are essentially parallel. Actually the light converges slightly because it has passed through a lens of long focal length just above the lower polarizing prism.

Plane-polarized light entering the crystal fragment is, in general, resolved into two components of different velocities vibrating in mutually perpendicular planes. The light resulting from the combination of these two components as they leave the crystal is again resolved in the analyzing prism, which permits light vibrating in only one plane to pass through.

If the path difference produced by the crystal plate for monochromatic light is one wave length or any whole number of wave lengths, the result between crossed polarizing prisms is darkness. If a fractional path difference is produced by the plate, the crystal plate is illuminated in all positions except the extinction positions. Maximum illumination is observed in the 45-degree positions when the path difference is $n\lambda/2$, and n is a whole odd number. In white light the interference color that is seen is dependent on the thickness, orientation, and birefringence of the crystal plate.

Under the conoscope the observed effects depend upon thickness and birefringence as under the orthoscope, but the orientation is in effect not constant for the light following the various rays through the crystal plate. Insertion of the accessory substage lens of short focal length produces a strongly convergent cone of light which passes through the plate in a great many directions. The path difference produced by the double refraction of light moving along one ray in the cone in general is not equal to the path difference resulting from the double refraction of light along another ray.

A crystal plate under the conoscope generally produces an interference figure consisting of curved color bands called *isochromatic curves,* and black or gray brushes called *isogyres.* In biaxial crystals,

as in uniaxial crystals, the appearance and behavior of the isochromatic curves and isogyres as the stage of the microscope is rotated depends to a large extent on the orientation of the crystal plate with respect to the light in the rays constituting the incident cone.

Isochromatic Curves. Isochromatic curves in biaxial crystals may be explained fully only by complex mathematical derivations. Emphasis is placed here on the more important visual concepts that lead to an elementary understanding of the nature of these curves.

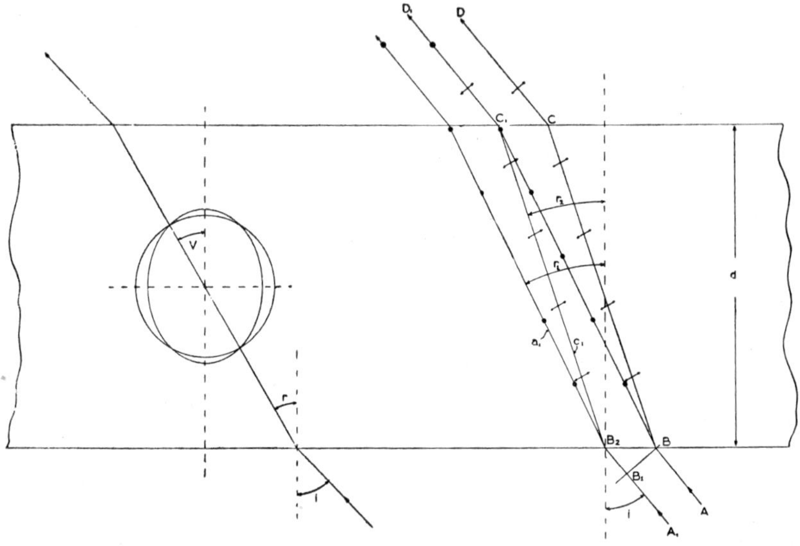

FIG. 1. Section in optic plane of biaxial crystal plate.

Figure 1 shows a cross section of a plate of a biaxial positive crystal. The plane of the drawing is parallel to the optic plane of a crystal with an optic angle of 60 degrees. The orientation of the ray velocity surfaces is indicated in Fig. 1*A*.

The effects of interior and exterior conical refraction are disregarded, and the primary and secondary optic axes are assumed to be coincident. Convergent light from the condensing lens incident on the lower surface of the plate in general is doubly refracted. However, if for the light in a given ray the angle of incidence is such that the angle of refraction is equal to V (half the optic angle), the light will travel along the optic axis with no double refraction. Inasmuch as all light travels along either optic axis with the same velocity and vibrates with equal ease in all directions, no path difference is produced, and

the result is darkness. Under the conoscope the points of emergence of the optic axes appear as dark spots.

A more general case is illustrated in Fig. 1*B*. Suppose that BB_1 is the wave front of a wave of monochromatic light incident on the lower surface of a biaxial crystal plate. Of all the rays transmitting light traveling normal to the wave front BB_1, two will be spaced so as to give the effects shown in the diagram. Consider AB the path of light along one of these rays. At B the light is doubly refracted and split into two components, BC_1 and BC, vibrating in mutually perpendicular planes; at emergence each component is again refracted and the components become parallel. Light moving along ray A_1B_2 is similarly affected. Upon emergence components following BC_1 and B_2C_1 travel the same path but are out of phase. That is, the crystal plate has produced a path difference between the two components.

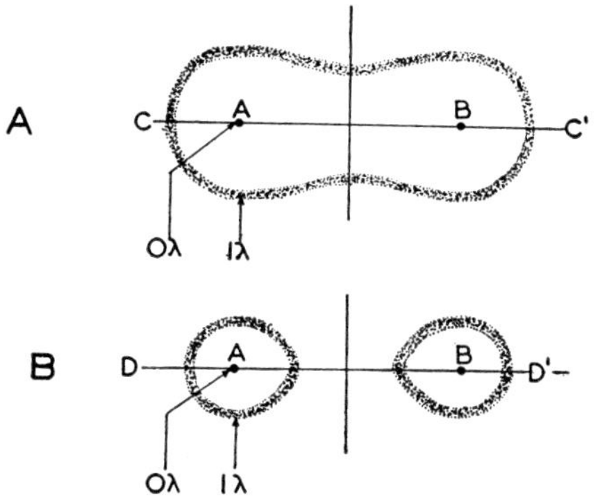

FIG. 2. Cassinian curves for path difference of 1λ. A. Thin plate. B. Thicker plate.

Under the conoscope in monochromatic light, the isochromatic curves appear as alternating dark and light curves. As in uniaxial crystals, an individual curve represents the *loci of the points of emergence of all components of light with the same path difference.* The dark curves mark the points of emergence of components having a path difference of $n\lambda$, where n is a whole number. The light curves indicate the points of emergence of components which have a path difference of $n\lambda/2$, where $n = 1, 3, 5, 7$, etc. However, the relationships in biaxial crystals are more complex than those in uniaxial crystals, for there are two optic axes instead of one, and three principal refractive indices instead of two.

Crystal plates cut normal to the acute bisectrix may be used to indicate the nature of isochromatic curves. Figure 2 shows dark curves about the points of emergence of the optic axes, A and B, of two crystal plates viewed in monochromatic light. The two crystal plates differ only in thickness. In order to avoid confusion, the isogyres are not shown.

In the thinner plate, Fig. 2*A*, the dark curve, sometimes called a *Cassinian curve,* outlines the points of emergence of components having

a path difference of one wave length. In the thicker plate, Fig. 2*B*, the dark curve of Fig. 2*A* has split into two almond-shaped curves, each of which lies about the point of emergence of an optic axis. As the thickness of a crystal plate changes, the dark curves for one wave-length path difference go through a series of changes which may be visualized by use of a three-dimensional surface called *Bertin's surface.*

Figure 3 shows such a surface for a path difference of 1 wave length for monochromatic light. The dark curves in Fig. 2 may be regarded

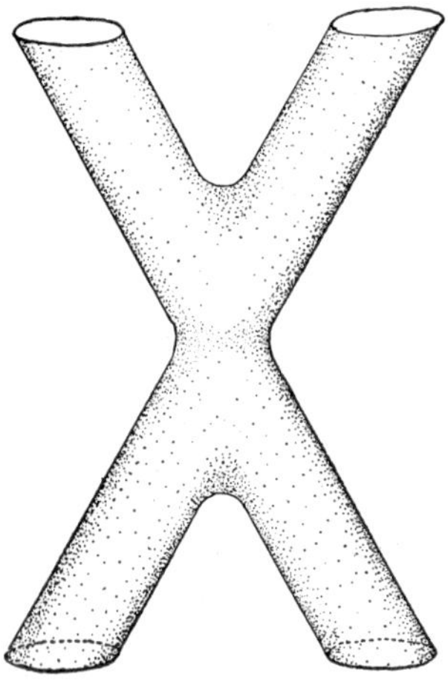

FIG. 3. Bertin's surface.

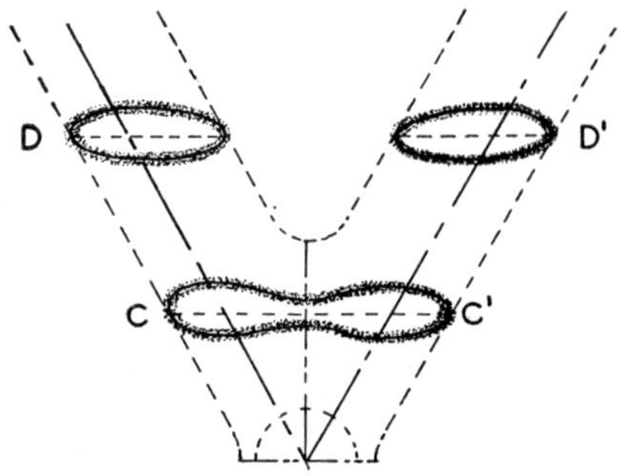

FIG. 4. Sections of Bertin's surface.

as cross sections of this surface occupying the relative positions indicated in Fig. 4. Bertin's surfaces, as such, are never seen in the conoscope, but isochromatic curves in interference figures may be regarded as cross sections of these surfaces.

Bertin's surfaces are surfaces of *equal path difference* and are very useful in predicting the nature of isochromatic curves for crystal plates with various orientations. These surfaces may be constructed for any path difference, and for a crystal viewed in monochromatic light the alternating dark and light curves may be considered as representing a section through a series of nested Bertin's surfaces. The light curves represent sections of surfaces of path difference $(n/2)\lambda$, $n = 1, 3, 5$, etc., and the dark curves correspond to surfaces of path difference $n\lambda$, $n = 1, 2, 3$, etc. Figure 5 shows the symmetry planes and Fig. 6 is a section through a series of nested Bertin's surfaces in a crystal plate cut at right angles to an acute bisectrix.

In interference figures viewed in white light the color curves repre-

sent sections of Bertin's surfaces of $(n/2)\lambda$ path difference for each color, where n is a whole odd number.

The number and spacing of the isochromatic curves permit estimation of the thickness or birefringence of a crystal fragment. The isochromatic curves are closely packed in interference figures of thick or highly birefringent crystals. Color curves are widely spaced or virtually absent in substances with low birefringences or in very thin plates of highly birefringent substances.

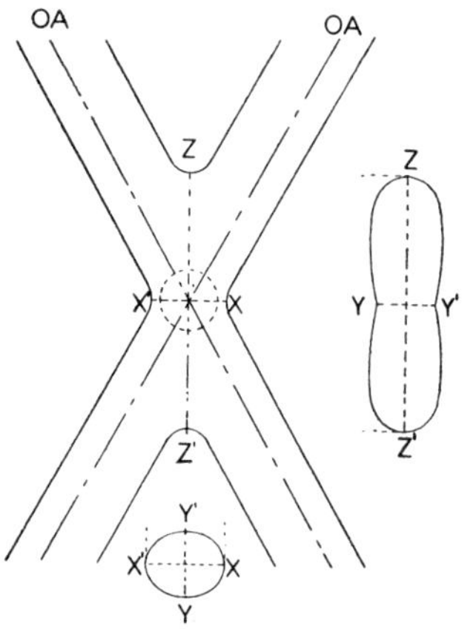

FIG. 5. Sections in symmetry planes of Bertin's surface.

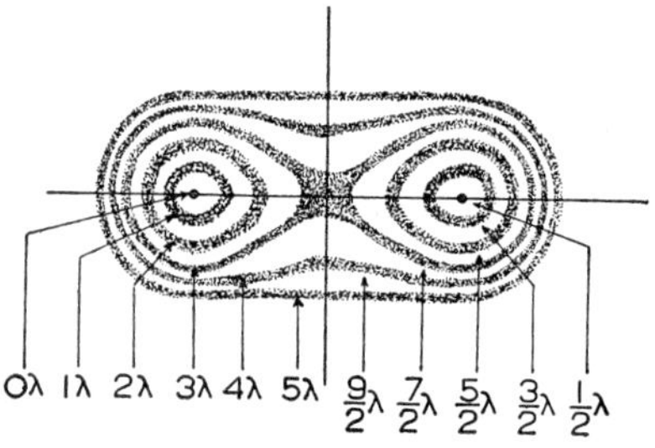

FIG. 6. Section through nested Bertin's surfaces. Section normal to acute bisectrix.

Isogyres. Under the orthoscope, between crossed polarizing prisms, a biaxial crystal plate extinguishes four times during a 360-degree rotation. The action of the conoscope is quite different. In interference figures of biaxial crystals, extinction produces black bars or brushes which change position as the stage of the microscope is rotated. These black areas are called isogyres, or *curves of equal vibration direction,* and appear in the form of crosses or hyperbolae. A complete interference figure consists of isogyres superimposed on isochromatic curves. Isogyres consist of curves determined by the *loci of points of emergence of light the traces of whose planes of vibration parallel or nearly parallel the planes of polarization of the upper and lower polarizing prisms.*

The explanation of isogyres is best approached by utilizing certain types of orthographic projections. The orthographic projection is especially useful in crystal optics because the interference figure seen in the conoscope is essentially an orthographic projection of interference phenomena in the focal plane of the objective lens system.

The strongly converging cone of light from the substage condenser

consists of light moving along a great number of rays, each bearing a different angular relationship to the optical directions of the crystal plate. In general, light in each ray, as it enters the crystal plate, is resolved into two components vibrating in mutually perpendicular planes. Each component has a velocity in the crystal plate which depends on several factors, but among all the components transmitted by the crystal plate there are some which have the same velocity despite the fact that they travel in different directions. If, instead of concerning ourselves with the passage of light along rays through the crystal, we turn our attention to the manner of passage of waves through the crystal, we will find a convenient mechanism for explaining isogyres. It should be remembered that a wave and the light in its corresponding ray vibrate in the same plane, the *vibration plane.*

If we ignore conical refraction, we may assume that monochromatic light waves travel with equal velocity along the optic axes of a biaxial indicatrix and vibrate in the circular section at right angles to the direction of propagation. Now, suppose that we know the direction and velocity of propagation along its wave normal of a particular wave of monochromatic light not vibrating in a circular section, and we wish to locate the positions of the wave normals of all other waves traveling with the same velocity. The velocity of a wave is inversely proportional to its refractive index, so that when we locate a curved section of the indicatrix containing radii equal to the radius giving the refractive index of the wave in question, we have located the vibration directions of a group of waves of equal velocity.

Figure 7 shows a curved section of the indicatrix containing equal radii. The wave normals corresponding to the radii constitute a cone, and the waves moving along the various wave normals in the cone have the same velocity.

The intersection of the cone and the surface of a sphere whose center is coincident with the center of the indicatrix produces a *spherical ellipse,* the center of which in Fig. 7 is the acute bisectrix, and the foci of which are the poles of the optic axes. A spherical ellipse is analogous to a plane ellipse in that the sum of the distances from the foci measured over the surface of the sphere to all points on the ellipse is constant. For each family of waves of constant velocity there is a corresponding spherical ellipse. Moreover, a second set of spherical ellipses with their centers at the pole of the obtuse bisectrix intersect the ellipses centered about the acute bisectrix. The foci of this second set of ellipses are the poles of the optic axes on either side of the obtuse bisectrix.

The two sets of ellipses on the surface of the sphere *intersect at*

right angles. Each ellipse gives the *loci of the points of emergence on the sphere of all waves of a given velocity.* By construction, the *trace of the plane of vibration* of the wave emerging at a particular point on a circular ellipse is perpendicular to the ellipse at the point of emergence. Thus, perpendiculars drawn to the ellipses at the point

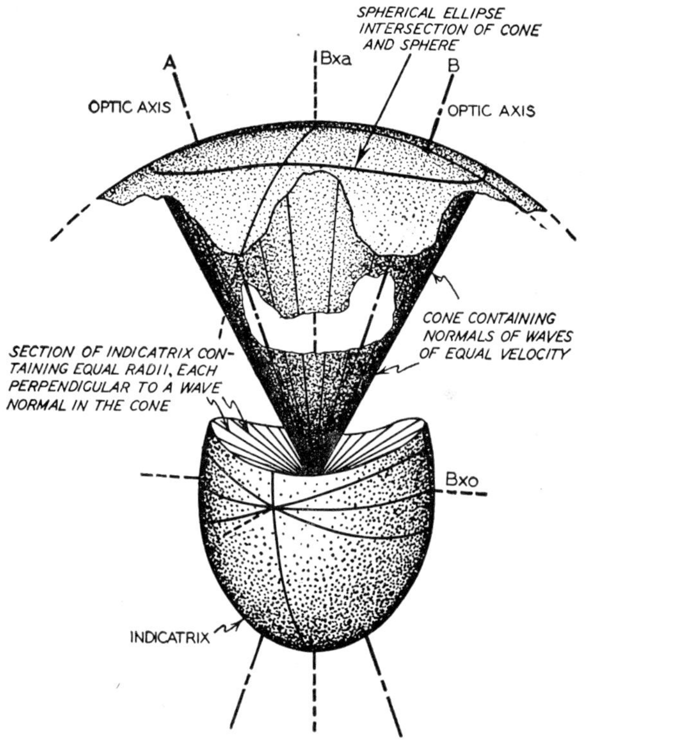

FIG. 7. Diagram showing origin of spherical ellipse about the points of emergence of the optic axes on a spherical projection.

of intersection of two ellipses give the traces of the mutually perpendicular vibration planes of the two components of the wave emerging at the intersection.

Moreover, inasmuch as a wave normal and its corresponding ray lie in the same plane, the plane of vibration, the perpendiculars to the ellipses at an intersection for all practical purposes give the traces of the vibration planes for the two components of light in a ray emerging at the intersection. Inasmuch as the ellipses intersect at right angles, the traces of the vibration planes also can be obtained by drawing tangents to the ellipses at a point of intersection.

Figure 8 portrays the relationships discussed above. *N* is the point of emergence of a wave on the surface of a sphere at the intersection of two circular ellipses. The perpendiculars (or tangents) to the two ellipses at the point of intersection give the traces of the vibration planes of the two components of this wave.

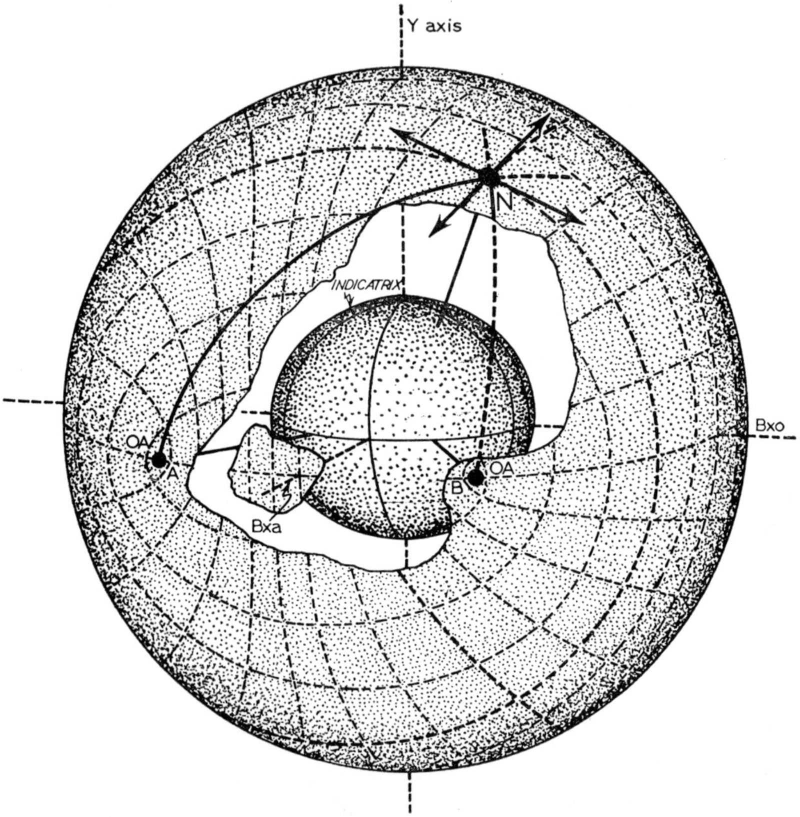

FIG. 8. Diagram showing relationship between an indicatrix and the spherical ellipses indicating the loci of the points of emergence of waves of equal velocity. Vibration directions for light emerging at a point *N* are indicated by arrows.

The spherical ellipses showing the loci of points of emergence of waves of equal velocity have been called *isotaques*. An orthographic projection of the isotaques into a diametral plane of the spherical projection is called a *skiodrome*.

Figures 9 to 11 are skiodromes of a biaxial negative crystal with a $2V$ of 60 degrees. In Fig. 9 the plane of projection is normal to X, the acute bisectrix. A and B are the points of emergence of the optic axes.

Figure 10 shows the skiodrome for the same crystal with the plane of the projection perpendicular to Z, the obtuse bisectrix. The plane of projection in Fig. 11 lies in the optic plane, the XZ plane.

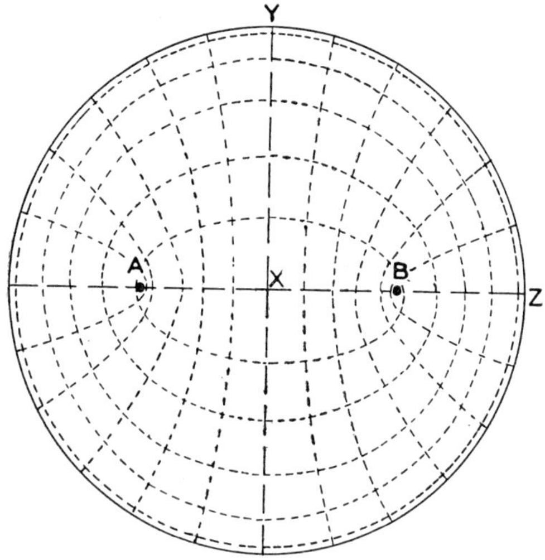

FIG. 9. Skiodrome of a biaxial crystal. Plane of projection YZ plane.

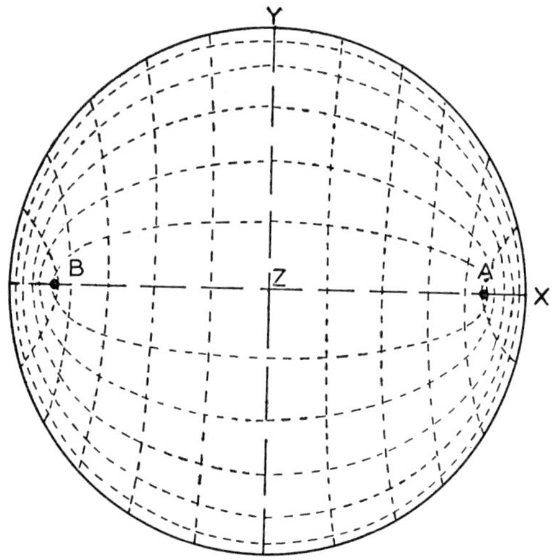

FIG. 10. Skiodrome of a biaxial crystal. Plane of projection XY plane.

The problem of determining the nature of the isogyres in an interference figure of a biaxial crystal plate in any position is solved by ascertaining the positions of emergence of light, the traces of whose vibration planes are parallel or nearly parallel to the planes of polariza-

tion of the upper and lower polarizing prisms. This is easily done on a skiodrome.

It will be noted that the curves of equal velocity do not intersect at right angles in the outer portion of the skiodrome. However, the

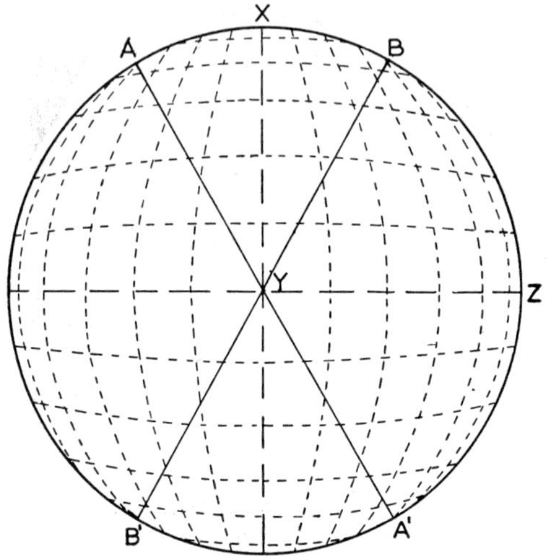

FIG. 11. Skiodrome of a biaxial crystal. Plane of projection XZ plane.

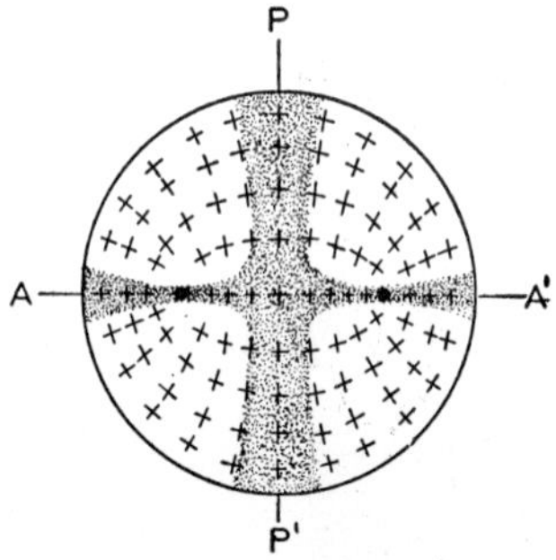

FIG. 12. Vibration directions obtained from skiodrome of acute bisectrix figure. Parallel position.

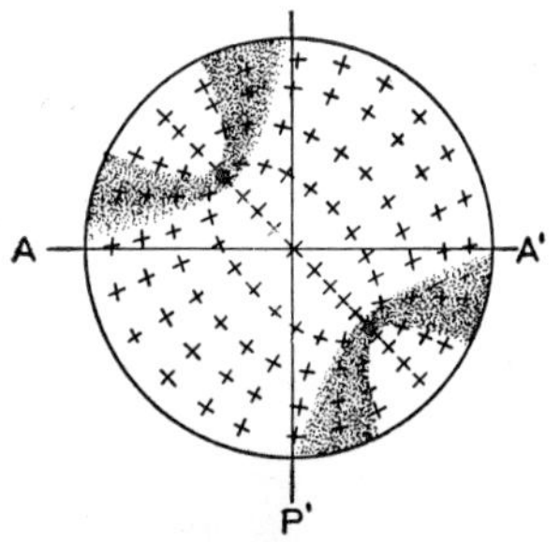

FIG. 13. Vibration directions obtained from skiodrome of acute bisectrix figure. Forty-five-degree position.

effect in the outer portion of the skiodrome may be disregarded because of the limited field of view of the conoscope.

Figures 12 and 13 illustrate the use of the skiodrome in predicting the positions of the isogyres. The traces of the vibration planes of the emergent light are obtained from a skiodrome normal to the acute bisectrix of a biaxial crystal.

In summary, a biaxial interference figure may be regarded as an interference effect in which isogyres are superimposed on isochromatic

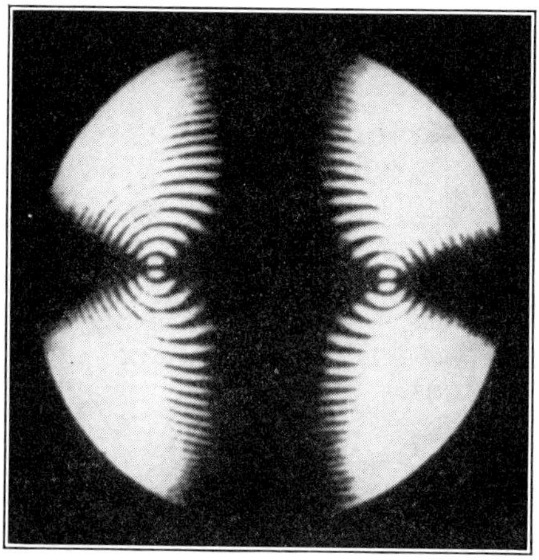

Courtesy McGraw-Hill Book Co.

FIG. 14. Acute bisectrix figure in parallel position. Monochromatic light. (Johannsen.)

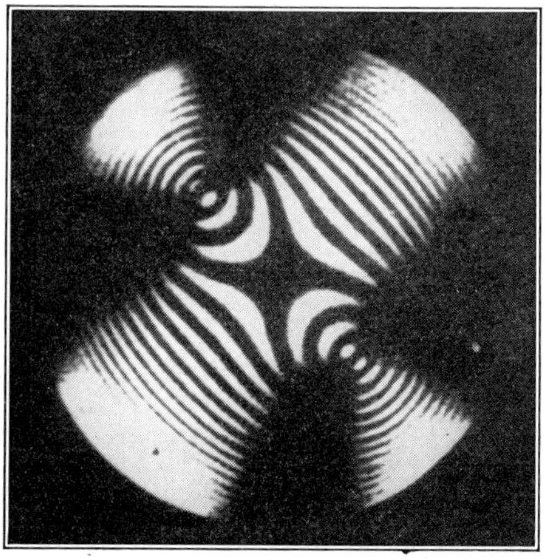

Courtesy McGraw-Hill Book Co.

FIG. 15. Acute bisectrix figure in forty-five-degree position. Monochromatic light. (Johannsen.)

curves. The isogyres are curves of *equal vibration direction* (with reference to the upper and lower polarizing prisms), and the isochromatic curves are *curves of equal path difference.*

The Acute Bisectrix Figure. A biaxial crystal section cut normal to the acute bisectrix gives, if the optic plane is parallel to either

the upper or lower polarizing prism, a black cross superimposed on color curves. As the stage of the microscope is rotated, the cross splits into two hyperbolae whose positions are determined by the optic angle ($2V$) and the position of the optic plane with reference to the polarizing prisms. This is illustrated in Figs. 14 and 15. (See also Figs. 12 and 13.) If the section is very thin or the birefringence is low, the color curves may not appear, and only the isogyre is seen.

Real and Apparent Optic Angles. The more acute angle between the optic axes is designated as $2V$, the optic angle. V is the angle between the acute bisectrix and either optic axis. In general, light traveling along the optic axis is refracted when it leaves the crystal and under the conoscope the points of emergence of the optic axis appear farther apart than they would if no refraction took place on emergence.

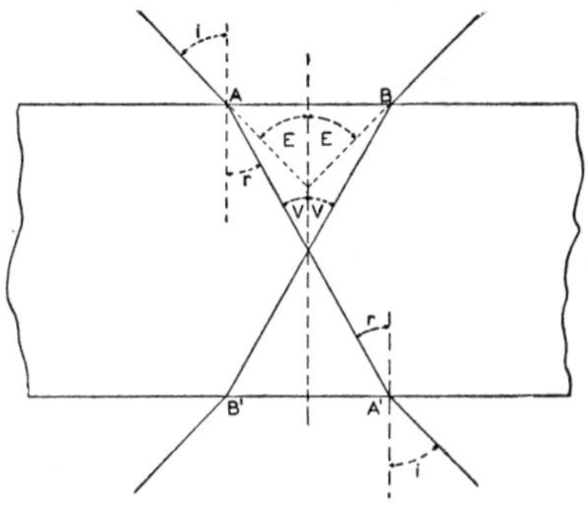

FIG. 16. Relationship of $2E$ to $2V$.

Refraction on emergence gives rise to an apparent optic angle, $2E$, which bears a simple relationship to $2V$. In Fig. 16, AA' and BB' indicate the directions of the optic axes in a crystal plate cut normal to the acute bisectrix. Light entering the crystal plate at A' with an angle of incidence i is refracted through an angle r. All light traveling path $A'A$ has an index equal to n_Y; so

$$\frac{\sin i}{\sin r} = n_Y$$

Inasmuch as

$$i = E$$

and

$$r = V$$

$$\frac{\sin i}{\sin r} = n_Y = \frac{\sin E}{\sin V}$$

or

$$\sin V = \frac{\sin E}{n_Y}$$

or

$$V = \sin^{-1} \frac{(\sin E)}{n_Y}$$

Figure 17 shows the relationship between $2V$ and $2E$ for various values of n_Y. The curves are a simple expression of the fact that light traveling along an optic axis is refracted less upon leaving crystals in which n_Y is low than it is upon leaving crystals for which the value of n_Y is higher.

The angle V may be calculated from the indices of refraction, but, if the indices are not accurately determined, considerable error is intro-

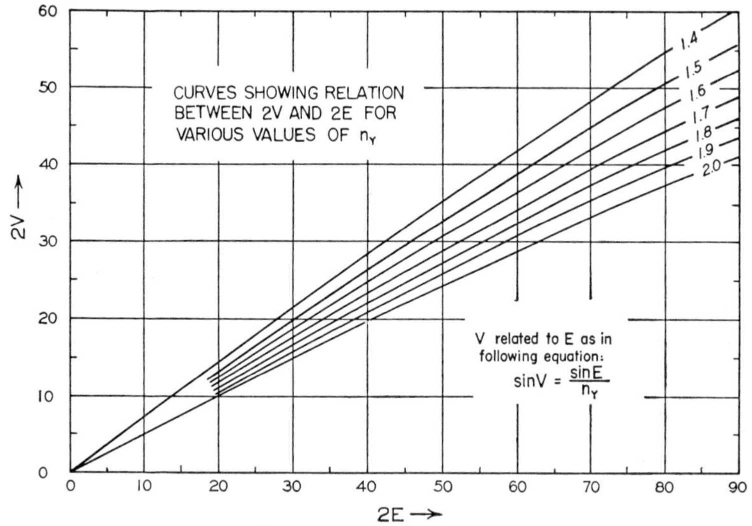

FIG. 17. Diagram showing relationship between $2V$ and $2E$ for various values of n_Y.

duced into the solution. If n_Y is known, it is possible to determine the optic angle of a mineral from an acute bisectrix figure by measuring the distance between the points of emergence of the optic axes with a micrometer eyepiece. Use is made of a value called *Mallard's constant.*

For a given combination of lenses the following equation is valid:

$$D = K \sin E$$

where D is half the scalar distance between the points of emergence of the optic axes as seen under the conoscope, K is Mallard's constant, and E is one-half the apparent axial angle in air. D is measured in terms of scale divisions in a micrometer eyepiece. Suppose that a crystal has a known $2E$ of 70 degrees and gives an acute bisectrix figure for which the points of emergence of the optic axes are 20 scale divisions of the micrometer eyepiece apart. Then

$$K = \frac{10}{\sin 35°} = 17.4 = \text{Mallard's constant}$$

Now

$$\frac{\sin E}{\sin V} = n_Y$$

So that for the particular combination of lenses used in the measurement

$$\sin V = \frac{D}{K \cdot n_Y} = \frac{D}{17.4 \cdot n_Y}$$

Crystals of pure ammonium sulfate serve conveniently for the determination of Mallard's constant for a microscope. The following optical data for ammonium sulfate are pertinent: n_Y for sodium light = 1.523; $2V = 52° 10'$; $2E = 84° 6'$.

Obtuse Bisectrix Figure. A crystal section cut normal to the obtuse bisectrix gives under the conoscope, when the optic plane is parallel to the vibration directions of either polarizing prism, an isogyre in the form of a black cross. But, because the points of emergence of the optic axes are generally outside of the field of view, slight rotation of the microscope stage causes the cross to break into two hyperbolae which leave the field very rapidly on further rotation. This is understood if reference is made to the skiodrome in Fig. 10.

The color curves in an obtuse bisectrix figure have the same general shape as those in an acute bisectrix figure, but ordinarily only segments are seen. These maintain the same shapes during rotation of the microscope stage.

Optic-Axis Figure. The optic-axis figure is seen in crystals cut normal to either optic axis. Properly oriented crystals under the petrographic microscope appear gray or black during a complete rotation of the microscope stage between crossed polarizing prisms, because virtually no path difference is produced in light traveling along an optic axis.

The optic-axis figure may be considered as one-half of an acute bisectrix figure; it consists of a single isogyre which goes through the same changes of shape and position as either of the segments of the acute bisectrix figure on rotation of the microscope stage. Figure 18 shows a photograph of an optic-axis figure in the 45-degree position.

Figure 19 shows diagrammatically the relationship between an acute bisectrix figure and an optic-axis figure. The observer is looking in a direction parallel to one of the optic axes, and, when the optic plane is parallel to either the upper or lower polarizing prism, the isogyre in the field of view is a straight bar. Figure 20 illustrates the effects of rotation of the microscope stage and demonstrates the reason for the change of the straight bar in the parallel position to a hyperbola in

the 45-degree position. Of special importance is the observation that *in the 45-degree position the convex side of the curved isogyre faces toward the acute bisectrix.*

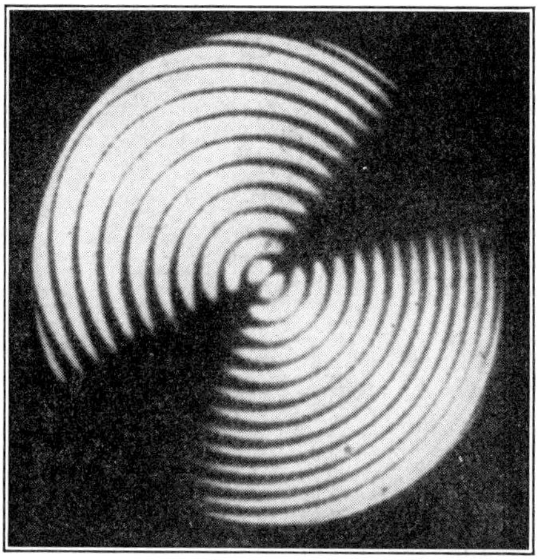

Courtesy McGraw-Hill Book Co.

FIG. 18. Optic-axis figure in 45-degree position. Monochromatic light. (Johannsen.)

Figures 19 and 20 are not true representations of actual conditions; for, if it were possible to see the whole area indicated in the drawing,

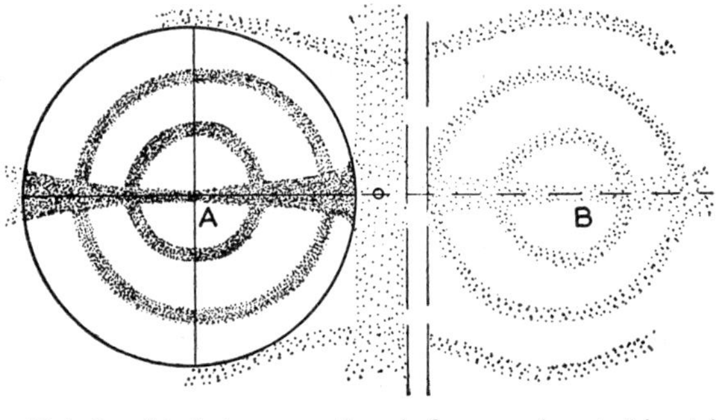

FIG. 19. Relationship between optic-axis figure and acute bisectrix figure. Parallel position.

the color curves around either optic axis would be highly asymmetrical with respect to each other.

Wright has shown that it is possible to estimate $2V$ from the curvature of the isogyre in an optic-axis figure in the 45-degree position. If the isogyre remains straight in all positions during rotation, the optic angle, $2V$, is 90 degrees. As the angle between the brushes of the

isogyre approaches 90 degrees, the optic angle approaches zero. The curvature of isogyres for several values of $2V$ are shown in Fig. 21.

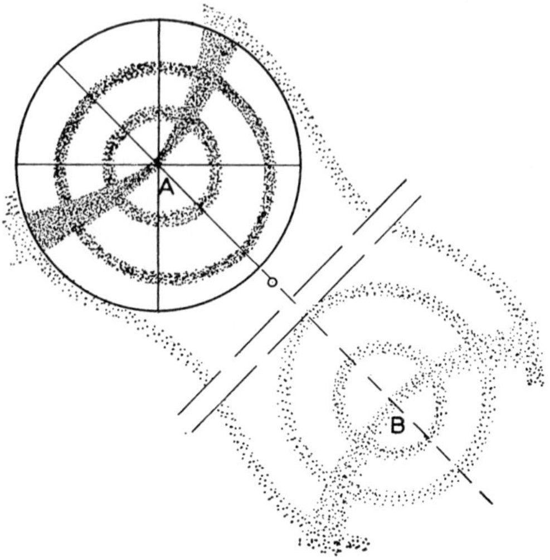

FIG. 20. Relationship between optic-axis figure and acute bisectrix figure. Forty-five-degree position.

These curves are based on calculations which assume a mean index of refraction of 1.60; hence they are not strictly accurate for crystals which have a mean index considerably lower or higher than 1.60.

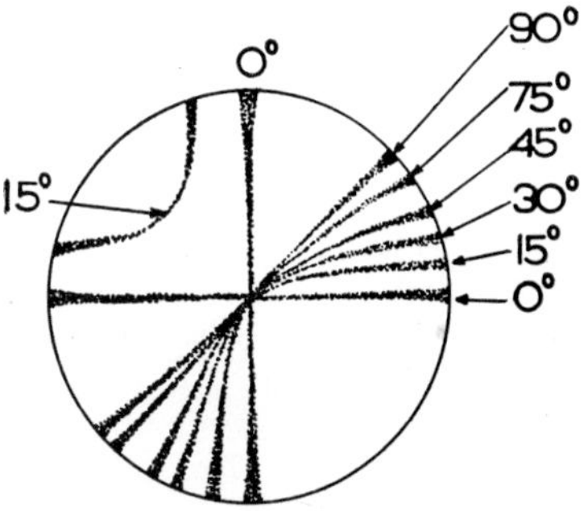

FIG. 21. Curvature of isogyres in optic-axis figures for various angles of $2V$. (After Wright.)

Optic Normal Figure. Sections cut perpendicular to the optic normal, the Y axis of the indicatrix, give broad, poorly defined isogyres, with or without color curves. When the X and Z directions parallel the traces of the vibration planes of the polarizing prisms, a diffuse cross appears which breaks into hyperbolae on slight rotation of the microscope stage. The hyperbolae leave the field rapidly as the stage is turned. The figure is similar to the flash figure seen in uniaxial crystals cut parallel to the c axis and is rarely used to determine optical sign.

The nature of the figure may be determined by referring to the skiodrome illustrated in Fig. 11 and to an XZ section of a Bertin's surface.

CHAPTER XV

DETERMINATION OF OPTIC SIGN IN BIAXIAL CRYSTALS

Introduction. Determination of optic sign in biaxial crystals may be made by various methods not involving the use of interference figures. It is possible to obtain sign by measuring the indices of refraction or by determining whether the optic angle, $2V$, lies about the X or Z axis. But sign determination from interference figures is relatively simple. In thin sections or in fragments in immersion media, an experienced observer can determine sign quickly and at the same time estimate the optic angle and the dispersion. Moreover, in interference figures, the orientation of the indicatrix may be ascertained, thus facilitating the accurate measurement of the refractive indices.

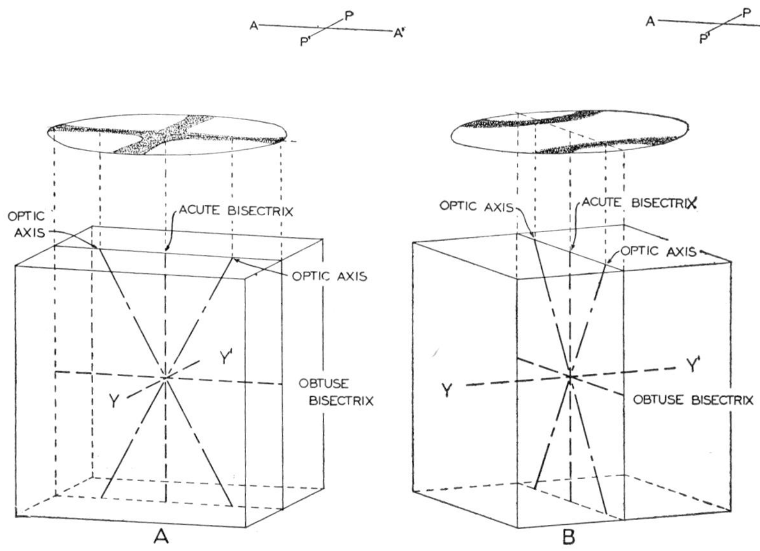

FIG. 1. Relation of acute bisectrix figure to indicatrix.

A. Parallel position. *B*. Forty-five-degree position.

Determination of Sign from Acute Bisectrix Figures. Figure 1 indicates the relationships between an acute bisectrix figure and the principal directions of the indicatrix. A centered acute bisectrix figure is seen under the conoscope in sections cut normal to the acute

bisectrix. Determination of optic sign consists essentially of ascertaining whether the Z axis or the X axis is the acute bisectrix. The optic sign may be determined in either the parallel or 45-degree position.

Of considerable value in sign determination is the *Biot-Fresnel law*, which may be employed to locate the traces of the planes of vibration of light following any particular path through a crystal. In Fig. 2, suppose that a random elliptical section has been cut through the center

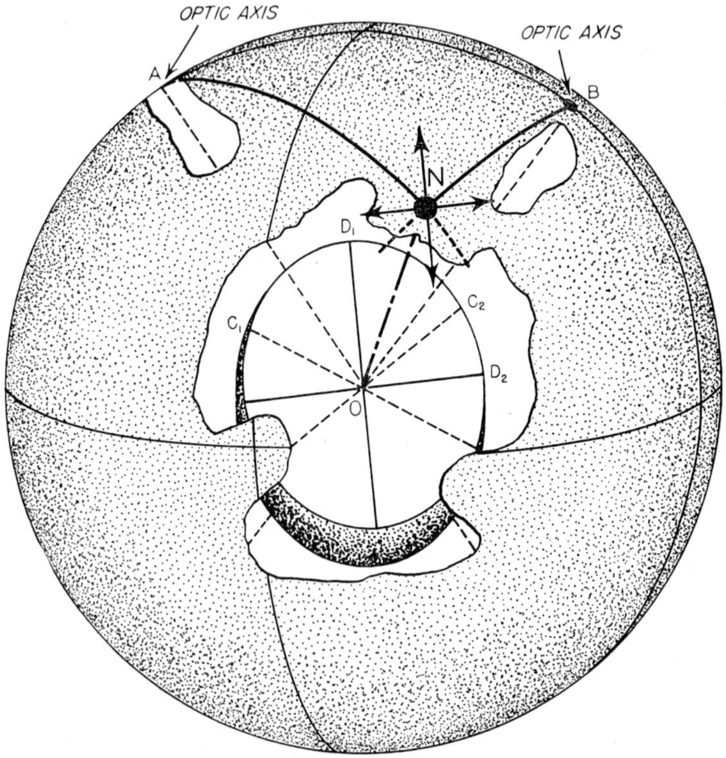

FIG. 2. Diagram used in proving Biot-Fresnel law.

of the indicatrix for a particular wave length of light, and that the indicatrix is enveloped by a sphere whose center coincides with the center of the indicatrix. ON is the direction of the wave normal perpendicular to the section of the indicatrix, and N is the point of emergence of the wave normal on the surface of the sphere. A and B are the points of emergence of the optic axes on the sphere, and AN and BN are arcs of great circles drawn from A and B to N.

According to the Biot-Fresnel law, the angles formed by lines joining the points of emergence of the optic axes and the point of emergence

of a wave on a section cut normal to ON, the wave normal, are bisected by the traces of the vibration planes of the two components of the wave moving along the normal. This relationship is illustrated in Fig. 2.

OC_1 and OC_2, two radii of the elliptical section of the indicatrix, are also the radii of the circular sections lying in the section of the indicatrix perpendicular to ON. Inasmuch as all radii of the circular sections are equal, OC_1 and OC_2 are equal, and, therefore, must lie symmetrically on either side of OD_1 and OD_2, the semimajor and semiminor axes of the elliptical section of the indicatrix. Now, the mu-

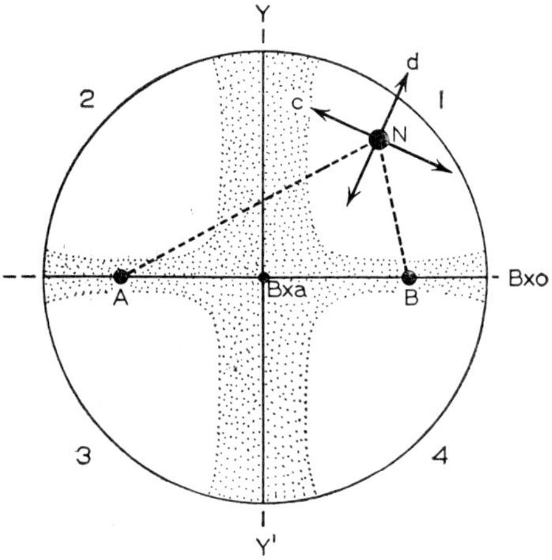

FIG. 3. Approximate method of locating traces of vibration planes of light emerging at any point N.

tually perpendicular vibration planes of the two components of the wave moving along ON include OD_1 and OD_2. Accordingly, the vibration planes bisect the angles between the planes including ON and the optic axes.

In an acute bisectrix figure the traces of the vibration planes of the two components of light emerging at any point N (Fig. 3) can be located approximately by connecting the points of emergence of the optic axes and the point of emergence of the light with straight lines. The directions bisecting the angles between the two lines at their intersection give the traces of the vibration planes of the two components.

The traces of the vibration planes can be obtained also from a skiodrome in which the plane of projection includes the obtuse bisectrix and the Y axis.

As an illustration of the use of the construction shown in Fig. 3, suppose that an accessory plate is inserted over the interference figure in such a manner that the fast direction of the plate is in quadrants 2 and 4. Suppose, moreover, that the insertion of the plate causes additive effects in quadrants 1 and 3. Light emerging at N consists of two components, the traces of the vibration planes of which are c and d. The vibration direction c is nearly parallel to the fast direction of the plate and, at the same time, is more nearly parallel to the vibration direction corresponding to the obtuse bisectrix (Bxo) than is component d. On the other hand, component d vibrates more nearly parallel to Y than does component c.

Inasmuch as the accessory plate demonstrates that component c is faster than component d, the component of light moving along the acute bisectrix and vibrating parallel to the obtuse bisectrix is faster than the component vibrating parallel to Y. Accordingly, Z is the acute bisectrix and the sign of the crystal is positive.

In negative crystals an accessory plate produces additive effects in quadrants 2 and 4 and subtractive effects in quadrants 1 and 3.

The Biot-Fresnel law can be used to obtain the traces of the vibration planes in an interference figure in any position of rotation of the microscope stage. However, in the 45-degree position the determination of sign may be made more easily by considering the effects of the accessory plates on light emerging along and near the trace of the optic plane.

Suppose that in an interference figure in the 45-degree position the trace of the optic plane is northwest-southeast, parallel to the fast direction of an accessory plate. Suppose, moreover, that the accessory plate produces additive effects between the isogyres (on the convex sides) and subtractive effects outside the isogyres (on the concave sides). Light moving along the acute bisectrix (Fig. 4) consists of two components vibrating at right angles to the direction of transmission and parallel to the obtuse bisectrix and the Y axis, respectively. If the effects are additive in the center of the figure, the component vibrating parallel to the fast direction of the accessory plate is faster than the component vibrating parallel to the Y axis. Therefore, the obtuse bisectrix is X and the acute bisectrix is Z, and the mineral is positive (Fig. 4A).

Now consider light following any path 1 in or near the optic plane and between the acute bisectrix and an optic axis. This light consists of two components: one vibrating parallel to Y, and the other vibrating normal to the direction of propagation and between the trace of the circular section and the obtuse bisectrix in the XZ plane. In positive

crystals the latter component has a lower refractive index than the Y component, and, accordingly, is faster. If the trace of the vibration plane of this component parallels the fast direction of the accessory plate, the effect at the point of emergence is additive. Using the same type of reasoning it can be shown that in positive crystals the effects

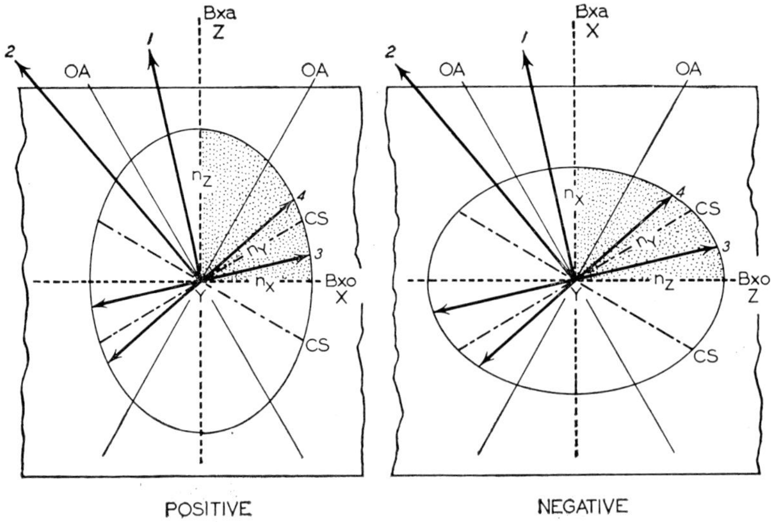

A. B

FIG. 4. Sections in optic planes of crystal plates cut normal to an acute bisectrix.

on the concave sides of the isogyres are subtractive if the fast direction of the accessory plate is parallel to the trace of the optic plane in an acute bisectrix figure.

If the optic plane of a positive crystal is perpendicular to the fast direction of the plate, the effects are the reverse of those described above.

Optic sign in acute bisectrix figures of negative crystals (Fig. 4*B*) is determined by demonstrating that X is the acute bisectrix.

Good results will be obtained if the mica plate and quartz wedge are used for interference figures displaying isochromatic curves, and the gypsum plate is used for diffuse figures with no color curves. The effects of the introduction of accessory plates over acute bisectrix figures in both the parallel and 45-degree positions are shown in Figs. 5, 6, and 7. All examples are optically positive; negative crystals give opposite results.

The observer rarely sees a perfectly centered acute bisectrix figure. Two conditions, or a combination of these, may arise: (1) The optic

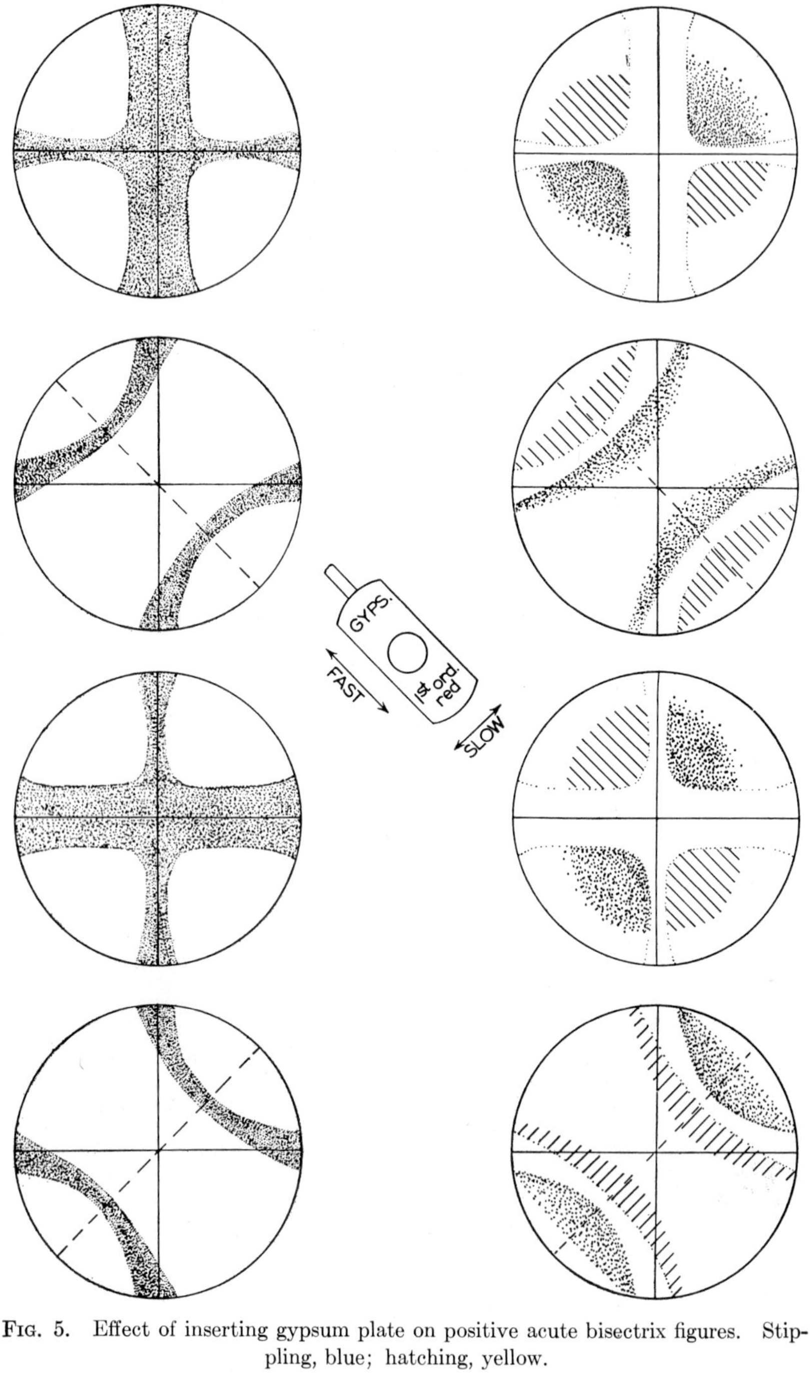

FIG. 5. Effect of inserting gypsum plate on positive acute bisectrix figures. Stippling, blue; hatching, yellow.

Fig. 6. Effect of insertion of mica plate on positive acute bisectrix figures.

plane may be normal to the plane of the section, but the acute bisectrix is inclined. This condition is illustrated in Fig. 8. (2) The

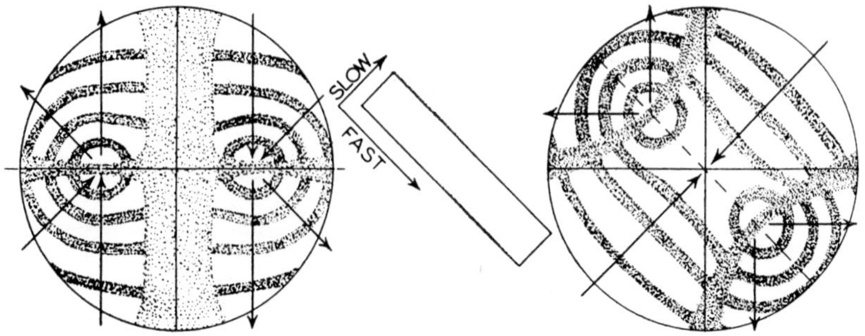

FIG. 7. Movement of color curves upon insertion of quartz wedge over positive acute bisectrix figures.

optic plane may be inclined, but the acute bisectrix and the optic normal both lie in a plane of symmetry between the isogyres. This

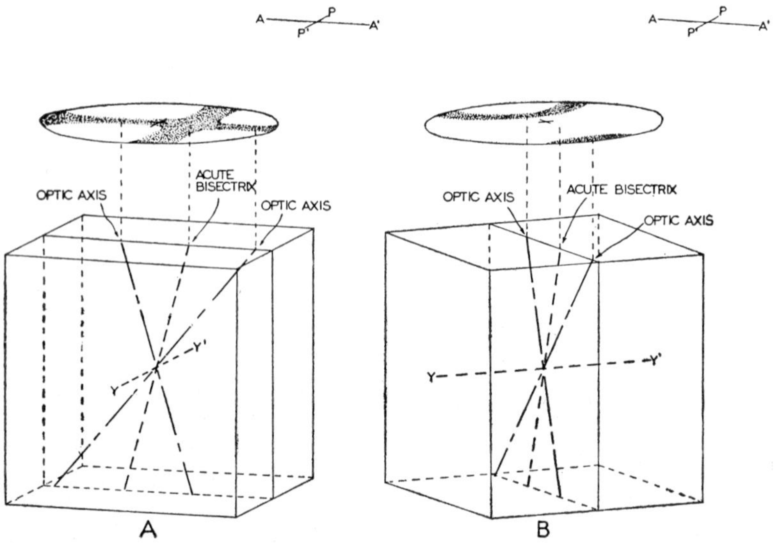

FIG. 8. Relation of off-center acute bisectrix figure to indicatrix.

A. Parallel position.
B. Forty-five-degree position.

condition is shown in Fig. 9. In general, however, both acute bisectrix and optic plane are inclined with respect to the section. The nature of an uncentered acute bisectrix figure of this type is illustrated in

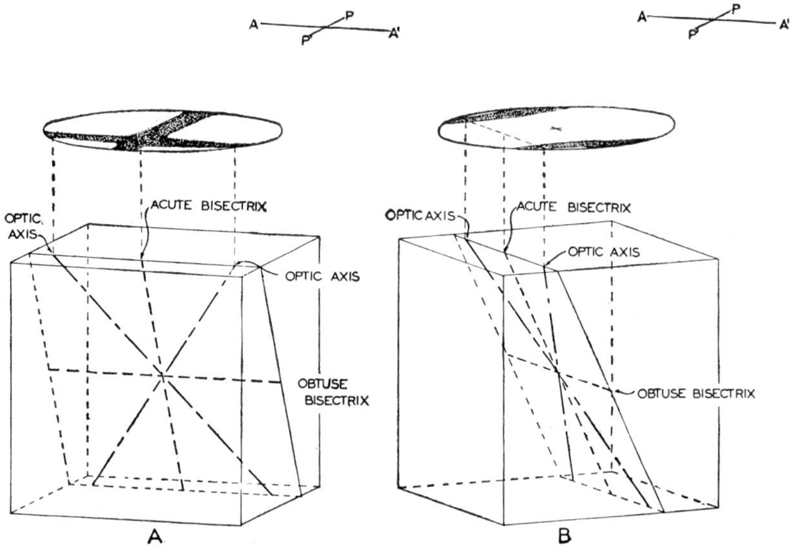

FIG. 9. Relation of off-center acute bisectrix figure to indicatrix.

A. Parallel position.
B. Forty-five-degree position.

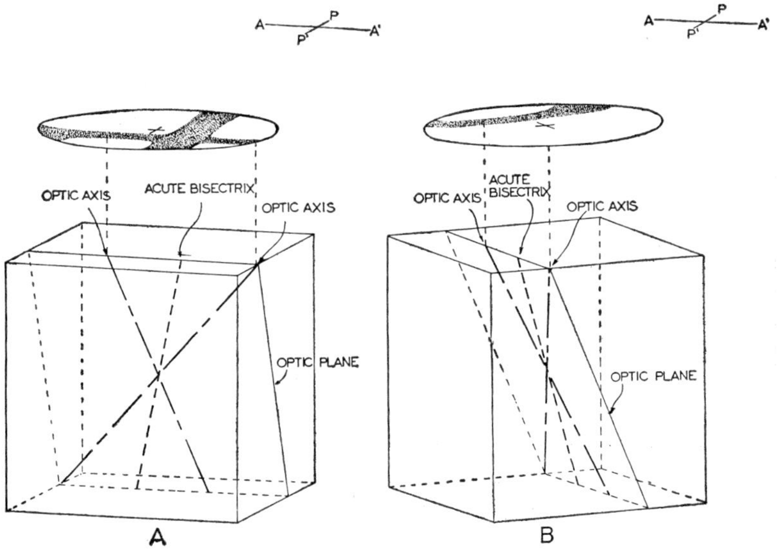

FIG. 10. General example of off-center acute bisectrix figure.

A. Parallel position.
B. Forty-five-degree position.

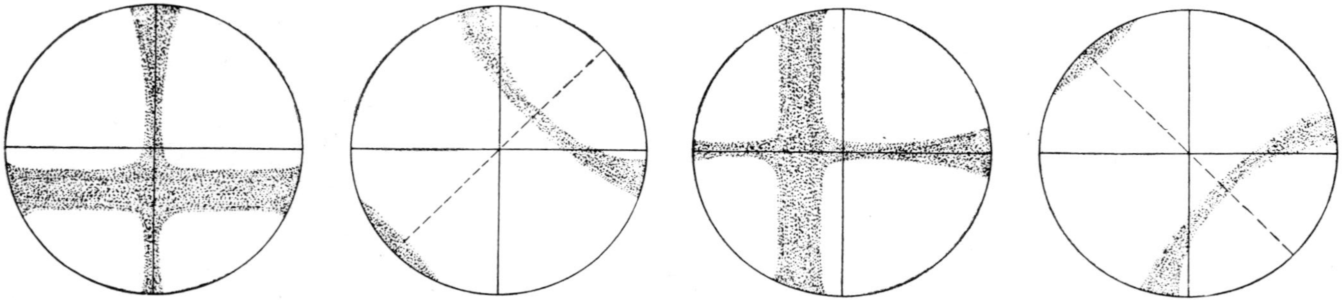

Fig. 11. Effect of clockwise rotation of microscope stage on a type of off-center acute bisectrix figure. (See Fig. 8.)

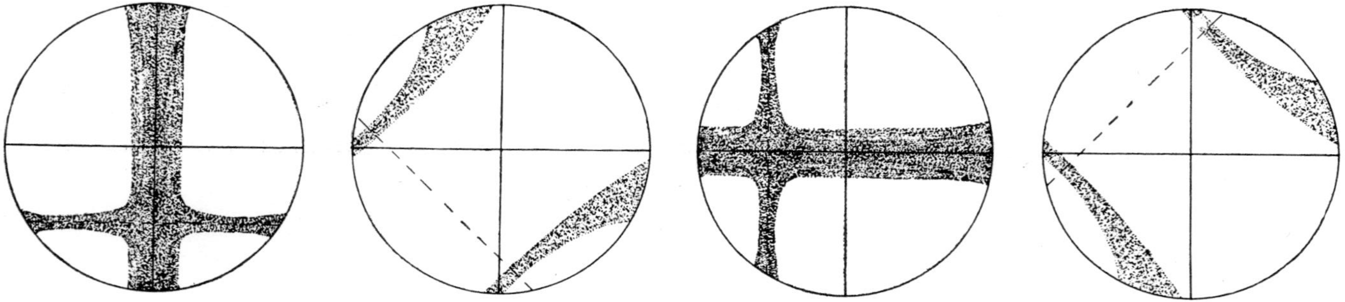

Fig. 12. Effect of clockwise rotation of microscope stage on a type of off-center acute bisectrix figure. (See Fig. 9.)

200   DETERMINATION OF OPTIC SIGN IN BIAXIAL CRYSTALS

Fig. 10. Figures 11 and 12 show the effects of rotation of the microscope on the isogyres of the two types of off-center acute bisectrix figures shown in Figs. 8 and 9.

If the position of the optic plane is determinable, acute bisectrix figures, even though considerably off center, may be used for sign determination.

Determination of Sign from Obtuse Bisectrix Figures. In obtuse bisectrix figures a black cross is observed in the zero and 90-degree

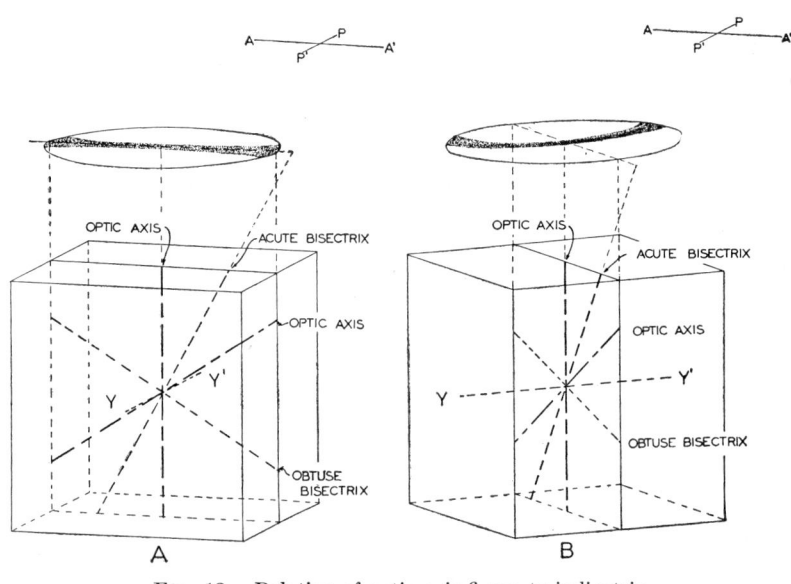

Fig. 13.   Relation of optic-axis figure to indicatrix.

A. Parallel position.
B. Forty-five-degree position.

positions. As the microscope stage is rotated, the cross rapidly breaks up into hyperbolae which leave the field in the quadrants containing the trace of the optic plane. If the observer recognizes the fact that he is looking along the obtuse bisectrix, rather than the acute bisectrix, the sign may be easily determined. But the effects obtained by introducing the accessory plates will be *opposite* from those obtained from an acute bisectrix figure. If $2V = 90$ degrees, there is no distinction between the acute and obtuse bisectrix figures.

Determination of Optic Sign from Optic-Axis Figures. The optic-axis figure, more than any other, is used to determine the sign of biaxial crystals because of the ease of finding crystal sections which give this figure. It is difficult, at times, to obtain a perfectly centered

OPTIC SIGN FROM OPTIC-AXIS FIGURES 201

figure, and it becomes necessary to use uncentered figures. The theory will be developed first for centered figures and then for uncentered figures.

Figures 13A and B illustrate the relation of the optic-axis figure to the indicatrix in a crystal plate. When the trace of the optic plane is parallel to the plane of polarization of the upper or lower polarizing prism, the isogyre is a straight bar, Fig. 13A. On rotating the crystal 45 degrees, a curved isogyre with the convex side toward the acute

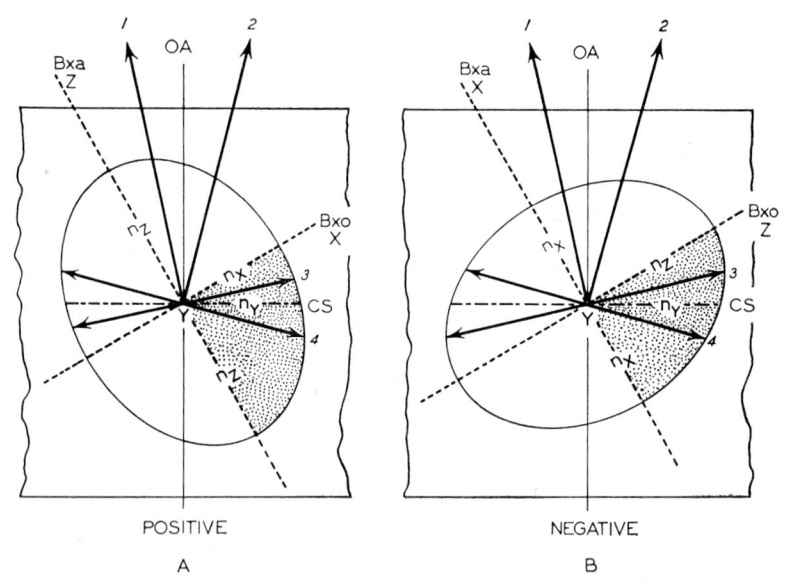

FIG. 14. Sections in optic planes of biaxial crystals cut normal to an optic axis.

bisectrix results, Fig. 13B. If 2V equals 90 degrees, the bar remains straight during a complete rotation of the microscope stage.

The interference figure in the 45-degree position permits the determination of sign. This is explained by reference to Fig. 14, which shows sections parallel to the optic planes of biaxial positive and biaxial negative crystals, both with an arbitrarily assigned 2V of 60 degrees.

In Fig. 14A suppose that light follows a path, 1, such that it emerges at a point lying in the trace of the optic plane between the points of emergence of the acute bisectrix and an optic axis. This light, then, emerges on the convex side of the isogyre and consists of two components: one vibrating parallel to the Y axis and having an index of n_Y, and the other vibrating in the optic plane at right angles to the direction of transmission and between a circular section and the obtuse

202   DETERMINATION OF OPTIC SIGN IN BIAXIAL CRYSTALS

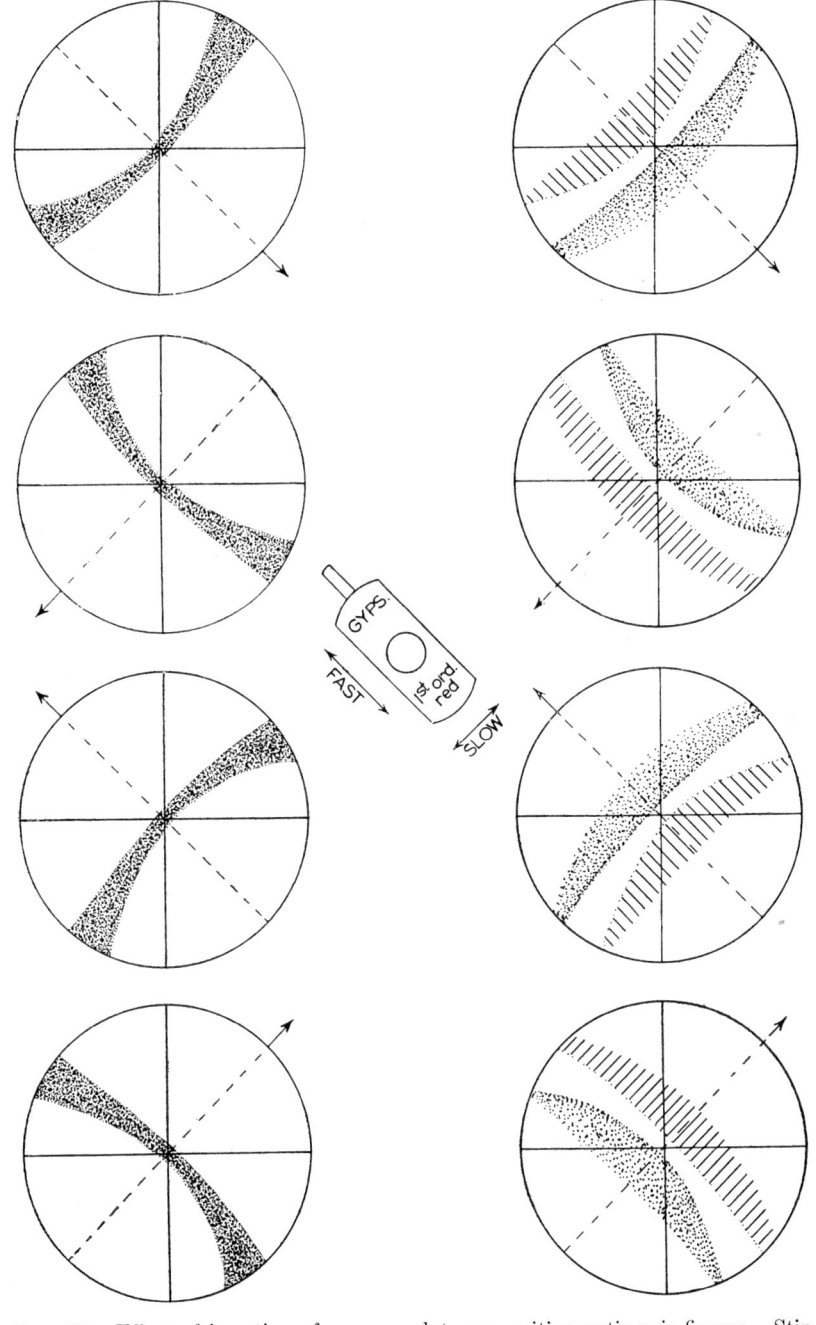

FIG. 15.   Effect of insertion of gypsum plate on positive optic-axis figures.   Stippling, blue; hatching, yellow.

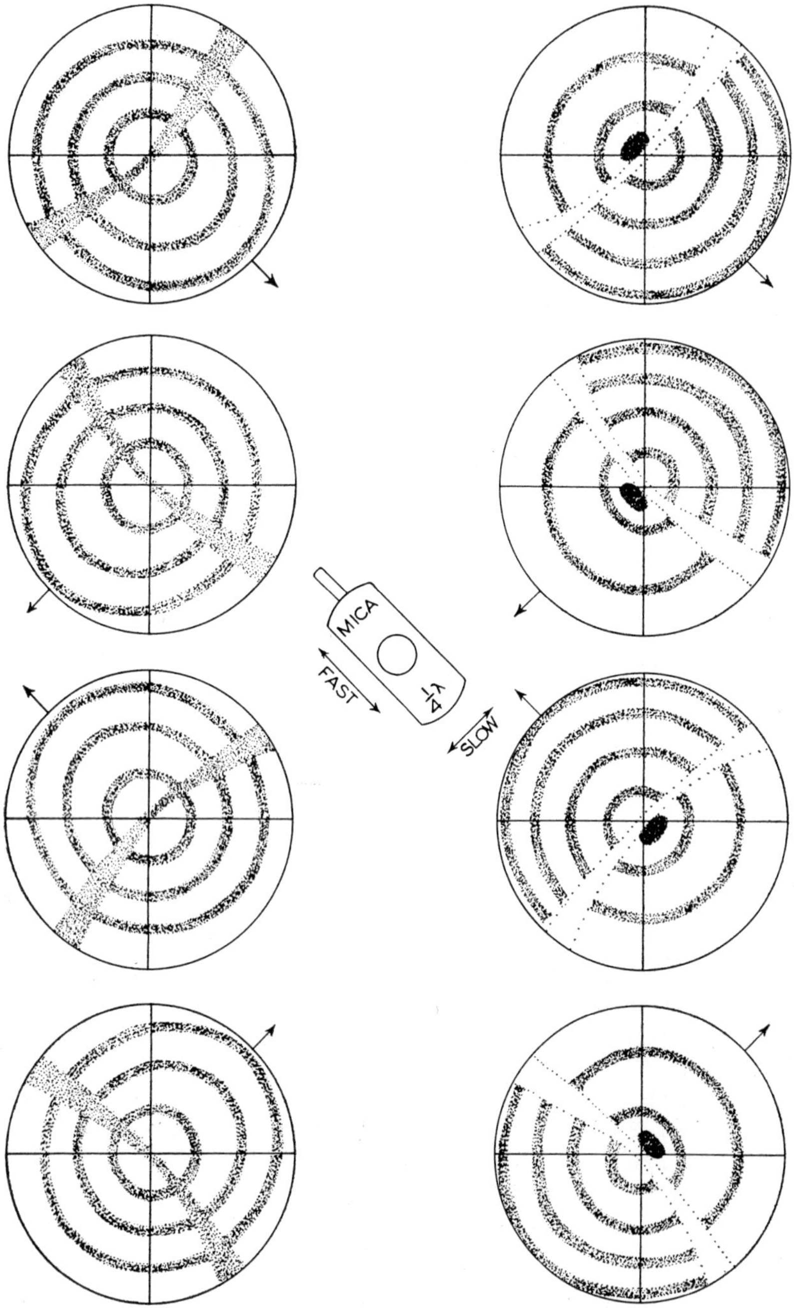

FIG. 16. Effect of insertion of mica plate on positive optic-axis figures.

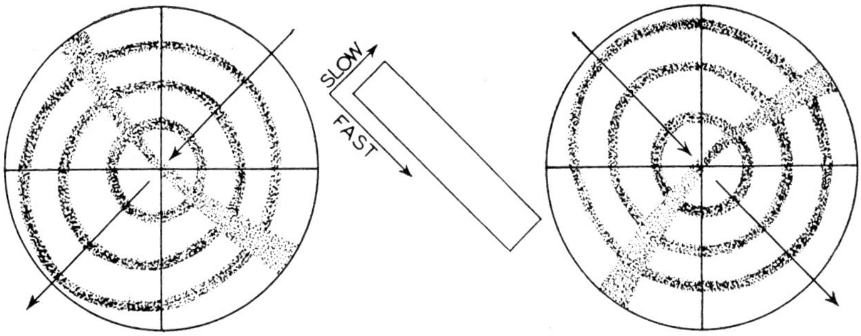

FIG. 17. Movement of color curves upon insertion of a quartz wedge over positive optic-axis figures.

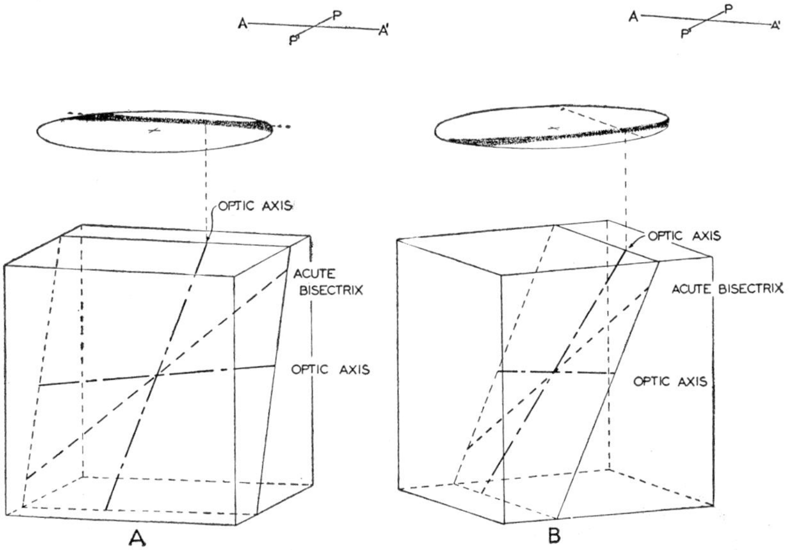

FIG. 18. Relation of off-center optic-axis figure to indicatrix.

A. Parallel position.
B. Forty-five-degree position.

bisectrix. If the crystal is positive, the component vibrating in the optic plane is faster than the component vibrating parallel to Y. Accordingly, if the fast direction of an accessory plate is parallel to the trace of the optic plane, an additive effect will result on the convex side of the isogyre, indicating that Z is the acute bisectrix. An opposite effect is seen on the convex side of the isogyre because the component in the optic plane for light emerging on the convex side of the isogyre vibrates between the acute bisectrix and a circular section.

Figure 14B shows the construction in the optic plane for a negative crystal.

In any event, the changes on the opposite sides of the isogyre upon insertion of an accessory plate will be opposite in nature. The specific changes depend on the position of the isogyre and the construction of the accessory plates.

Figures 15, 16, and 17 illustrate the changes that take place in optic-axis figures when various accessory plates are inserted into the microscope. All examples are positive; opposite effects will be noted in negative crystals.

In the usual type of off-center optic-axis figure, the optic plane as well as the optic axis is inclined with respect to the surface of the section, as shown in Fig. 18. This drawing permits visualization of the changes that take place during rotation of the crystal. If it is possible to locate the position of the trace of the optic plane and the approximate position of the acute bisectrix, the problem of sign determination becomes no more difficult than the determination of sign in centered figures. The effects on introduction of the accessory plates are similar, whether the figure is centered or uncentered.

In Fig. 19 various possibilities are indicated. A study of these possibilities together with a visualization of the relations suggested in Fig. 18 permits the solution of most problems of sign determination in optic-axis figures.

Use of Interference Figures in Index Measurement. Interference figures reveal the positions of the axes of the indicatrix and assist in measurement of the principal indices of refraction. In an acute bisectrix figure it is known that the Y direction is perpendicular to the optic plane, which includes the points of emergence of the optic axis. Moreover, when the optic sign is ascertained, the position of the X or Z axis is discovered. For example, in a positive crystal the two components of light traveling parallel to the acute bisectrix and vibrating in mutually perpendicular planes have indices equal to n_X and n_Y; n_Z may be measured in a grain which shows an obtuse bisectrix figure. Crystals which give negative acute bisectrix figures yield n_Y and n_Z. Centered optic-axis figures yield n_Y in all positions.

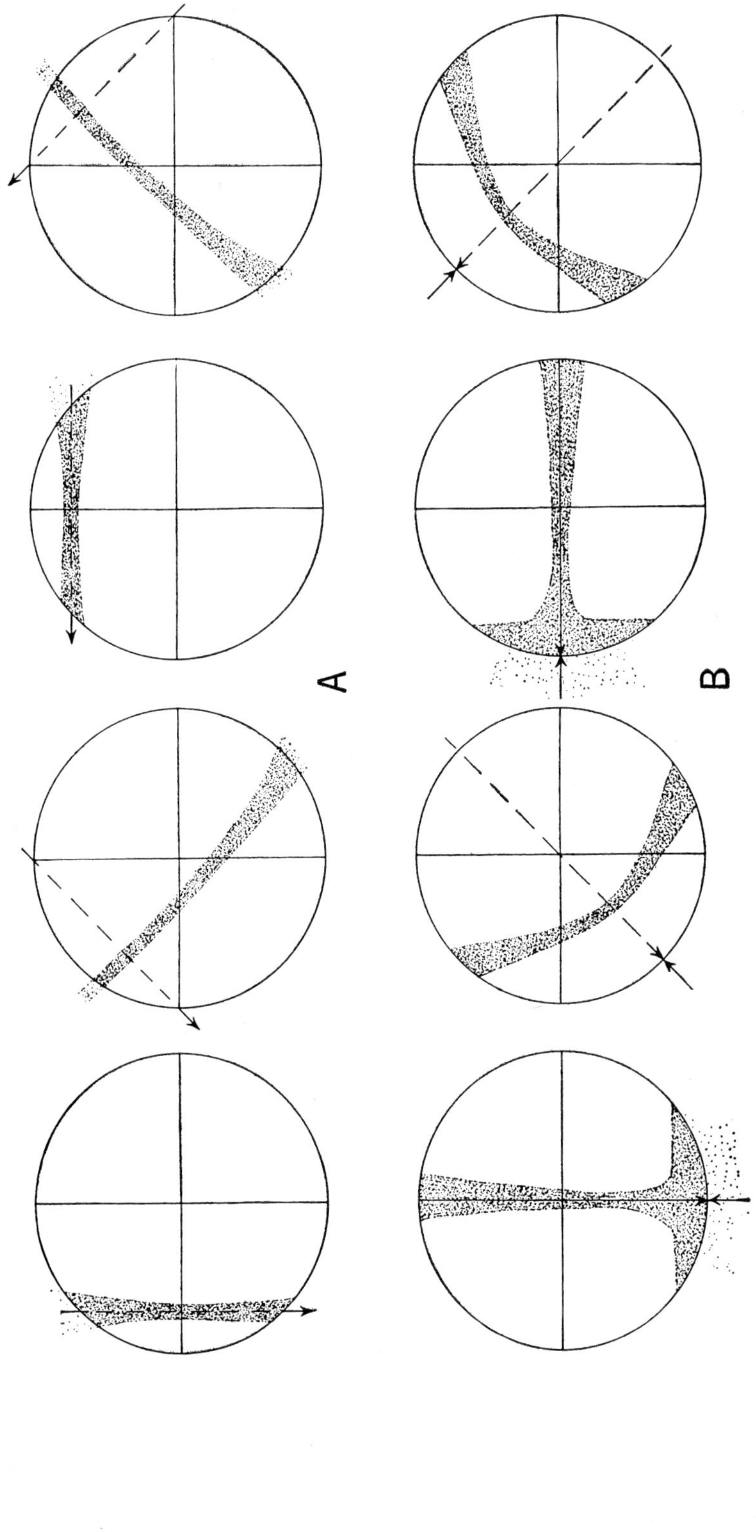
A
B

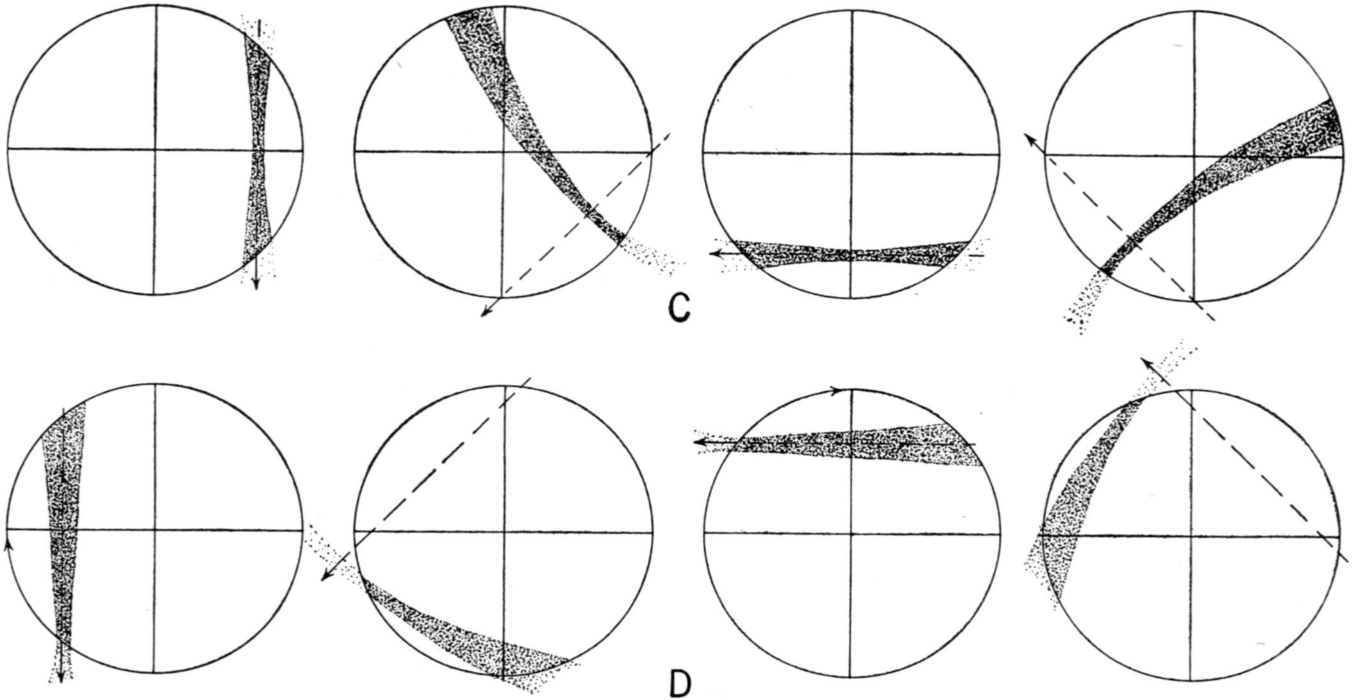

FIG. 19. Effect of clockwise rotation of microscope stage on various types of off-center optic-axis figures. Dashed line is trace of optic plane. Arrows point toward acute bisectrix.

Chapter XVI

DISPERSION IN BIAXIAL CRYSTALS

Introduction. *Dispersion* is an expression of the fact that the refractive indices of nonopaque substances vary with the wave length of light and the fact that passage of light through crystals conforms rigidly to the requirements of crystal symmetry.

In biaxial crystals all three principal refractive indices undergo dispersion resulting in dispersive effects involving the optic angle. Moreover, there are additional dispersive effects which depend on optic orientation as it is related to crystal symmetry. Dispersion introduces certain peculiarities into interference figures of biaxial crystals observed in white light. These peculiarities are attributable to the fact that for each wave length of light there exists an indicatrix which has its own characteristic refractive indices, optic angle, and optic orientation.

The most obvious manifestation of dispersion in biaxial crystals is the presence of color fringes on the isogyres of interference figures. These fringes may be pronounced or barely visible, depending upon whether the dispersion is strong or weak. The amount of dispersion as seen in interference figures may be expressed as follows: *perceptible* if the isogyres show faintly visible colored borders; *weak* if a little more easily seen; *strong* if very apparent; and *extreme* if the color fringes cover a large part of the field of the microscope.

Dispersion of the indices of refraction in biaxial crystals results in *dispersion of the optic axes.* The reason for this is apparent when it is considered that the optic angle, $2V$, may be computed from the indices. Accordingly, if the indices for two wave lengths of light are different, the optic angles differ. Dispersion of the optic axes in biaxial crystals is expressed by a formula which states whether the optic angle for red light is greater or less than the optic angle for violet light. These colors are chosen because they lie at the extremes of the spectrum and because they are the colors commonly seen in the color fringes. If $2V$ for red is greater than $2V$ for violet, the dispersion formula is $r > v$; $r < v$ expresses the reverse relationship. The violet is actually a bluish violet and sometimes almost a pure blue.

In addition to dispersion of the indices and $2V$, there are several

types of dispersion which depend on crystal symmetry. Recognition of the type of dispersion in a crystal serves to classify the crystal in its proper crystal system. The distribution in interference figures of color effects due to dispersion conforms rigidly to the symmetry requirements of each crystal system. Dispersion dependent on crystal symmetry in the monoclinic and triclinic systems is commonly designated as *dispersion of the bisectrices.*

Dispersion in Orthorhombic Crystals. In orthorhombic crystals the X, Y, and Z axes of the indicatrices for all wave lengths of light coincide with the crystallographic axes. This results from the symmetry requirements of the orthorhombic system, which, in the dipyramidal class, has three mutually perpendicular planes of symmetry. Dispersion of the indices, that is, variation of the indices with the wave length of light, causes variation in the optic angle, $2V$. This type of dispersion is called *rhombic* or, perhaps better, *orthorhombic dispersion.* It is the principal type of dispersion observed in interference figures of orthorhombic crystals.

The origin and nature of rhombic dispersion is suggested by Figs. 1 and 2, which show the dispersion effects produced in an acute bisectrix figure. Dispersion of the indices has caused dispersion of the optic axes for all wave lengths of light, but the figures show only the positions of the optic axes of red and violet light. Arbitrarily it is assumed that the optic angle, $2V$, for red is greater than the optic angle for violet; that is, $r > v$.

In the zero position, Fig. 1*A*, the points of emergence of the optic axes for all colors lie in the trace of the optic plane, and no color effects are observed. However, in the 45-degree position, Fig. 1*B*, color fringes appear on the concave and convex sides of the hyperbolic isogyres. The color fringes are symmetrically disposed with respect to the isogyres and the optic plane.

The fact that violet (or blue) fringes appear at the points of emergence of the optic axes for red, and red fringes appear at the points of emergence of the optic axes for violet, needs explanation. This phenomenon is understood if it is realized that no path difference is produced for red light moving along an optic axis for red. Consequently the red light, in effect, is removed from the spectrum at the points of emergence of its optic axes, and a color near the other end of the spectrum appears, that is, blue. The same type of reasoning applies to violet light. That is, red light appears at the points of emergence of the optic axes for violet light.

Figure 2 shows orthorhombic dispersion in plan. The isochromatic curves are omitted because, although they are commonly present in

210 DISPERSION IN BIAXIAL CRYSTALS

interference figures showing dispersion, their representation in the drawings does not add to the understanding of the principles here discussed.

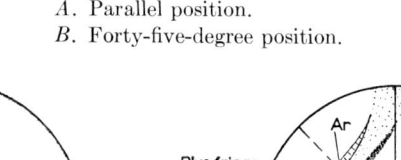

Fig. 1. Rhombic dispersion in an orthorhombic crystal: $r > v$.

A. Parallel position.
B. Forty-five-degree position.

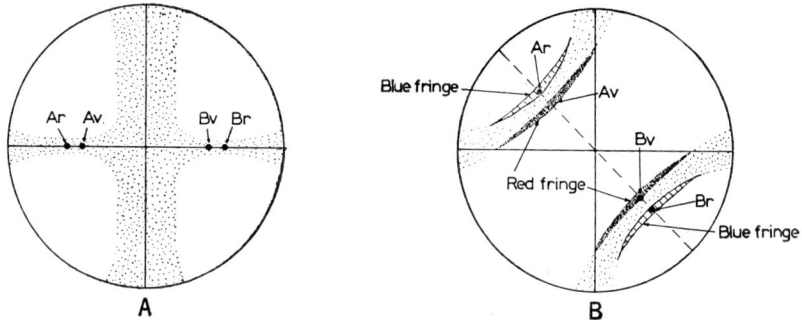

Fig. 2. Rhombic dispersion in plane: $r > v$.

A. Parallel position.
B. Forty-five-degree position.

Between the red and blue fringes a normal black or gray isogyre is present where the path differences of the intermediate colors of the spectrum are insufficient to produce color effects.

In crystals in which $r < v$, the color fringes are reversed.

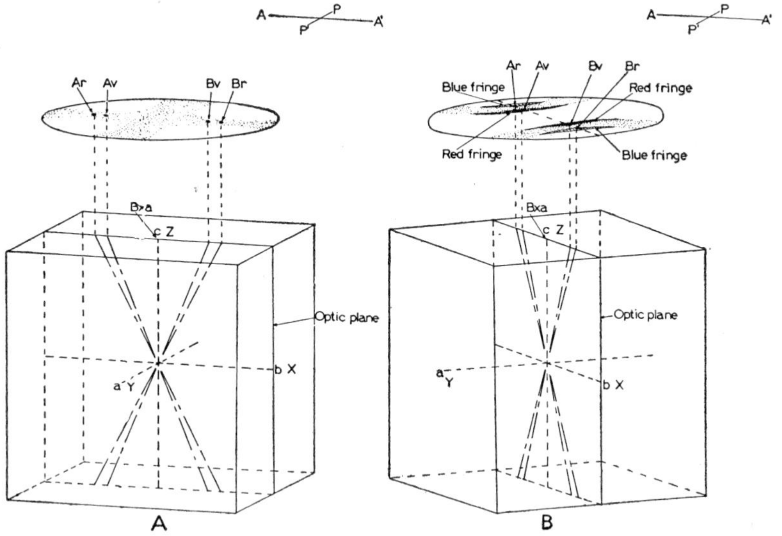

DISPERSION IN ORTHORHOMBIC CRYSTALS   211

Figure 3 is a three-dimensional drawing of an orthorhombic crystal cut away to show the indicatrices for red and blue (violet). Note that the optic orientation for both indicatrices is the same.

In certain rare orthorhombic crystals, such as brookite, the dispersion is so strong that a very unusual effect is produced. For red ($\lambda = 670$ mμ) Z is the acute bisectrix, and the optic plane lies in the

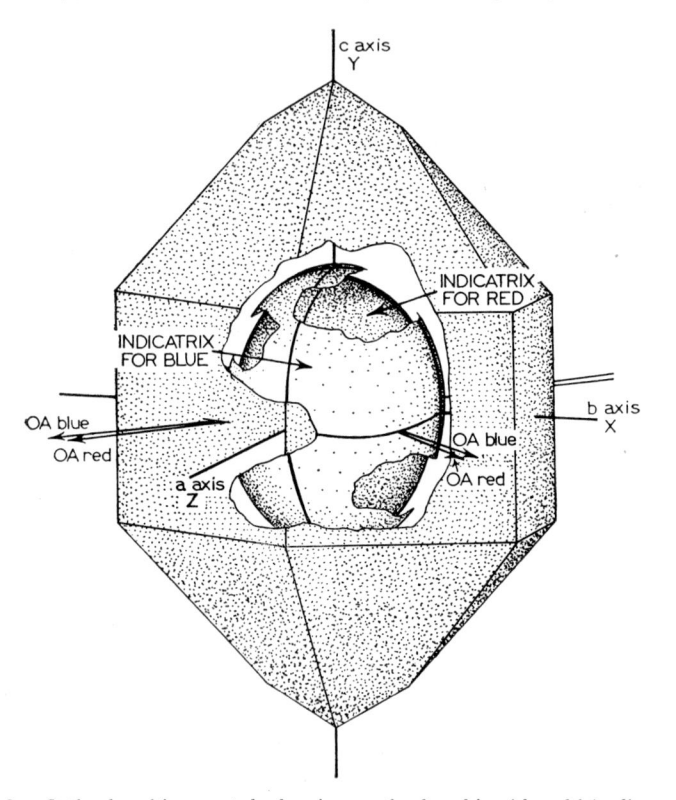

FIG. 3. Orthorhombic crystal showing orthorhombic (rhombic) dispersion. Crystal is positive and $2V$ is near 90 degrees.

ab plane of the crystal. As the wave length is decreased, the optic angle decreases and becomes zero at a wave length of 555 mμ. For wave lengths less than 555 mμ, the optic axis opens up in the ac plane, and X becomes the acute bisectrix. This type of dispersion is called *crossed axial plane dispersion* and is indicated digrammatically in Fig. 4, a drawing of a crystal of brookite.

In describing the optic orientation of crystals such as brookite, it is necessary to specify the wave length of the light used in the determination.

212   DISPERSION IN BIAXIAL CRYSTALS

Dispersion in Monoclinic Crystals. Monoclinic crystals in the prismatic class have a single plane of symmetry, which includes the a and c crystal axes; the b crystallographic axis is normal to the symmetry plane. Because of the symmetry requirements of monoclinic crystals, one axis of the indicatrix for each wave length of light must coincide

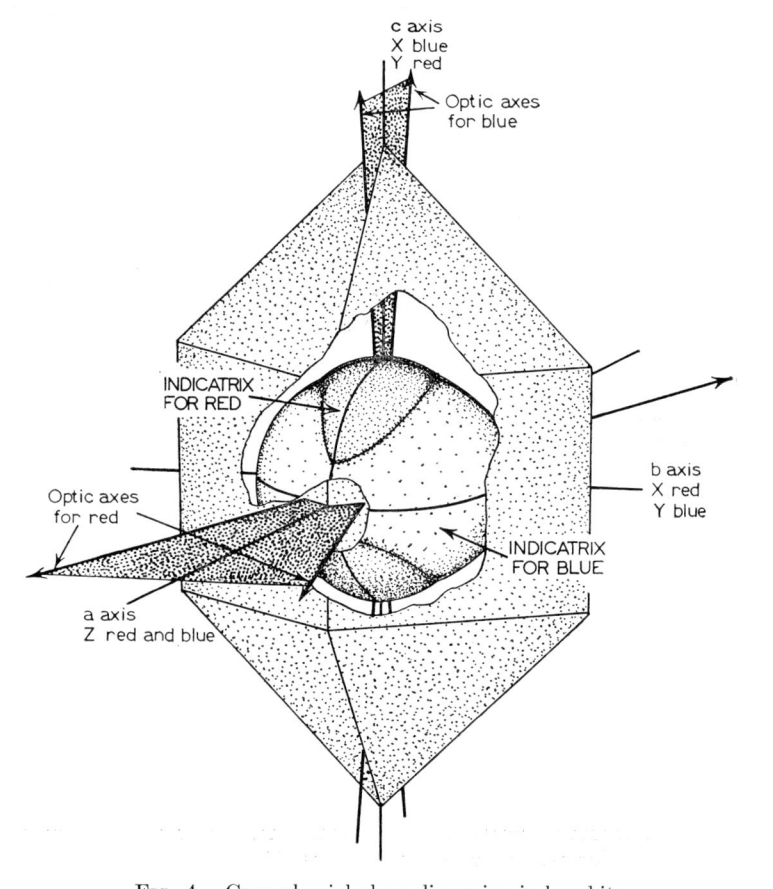

FIG. 4. Crossed axial plane dispersion in brookite.

with the b crystallographic axis, and the other axes of the indicatrix must lie in the crystallographic symmetry plane. Three possibilities arise: (1) The acute bisectrix parallels the b axis, and the obtuse bisectrix and the optic normal lie in the crystallographic plane of symmetry. The b axis serves, in effect, as an axis of rotation for the indicatrices of the various colors. In this example interference figures in white light show *crossed dispersion* in an acute bisectrix figure. (2) The obtuse bisectrix parallels the b axis, and the acute bisectrix

and the optic normal lie in the crystallographic plane of symmetry. This gives rise to *horizontal dispersion* in the acute bisectrix figure observed in white light. (3) The optic normal parallels the b axis, and the acute and obtuse bisectrices lie in the crystallographic plane of symmetry. Acute bisectrix figures show *inclined dispersion* in white light.

Figures 5 and 6 show the space relationships that result in *crossed dispersion.* This phenomenon is seen in monoclinic crystal plates

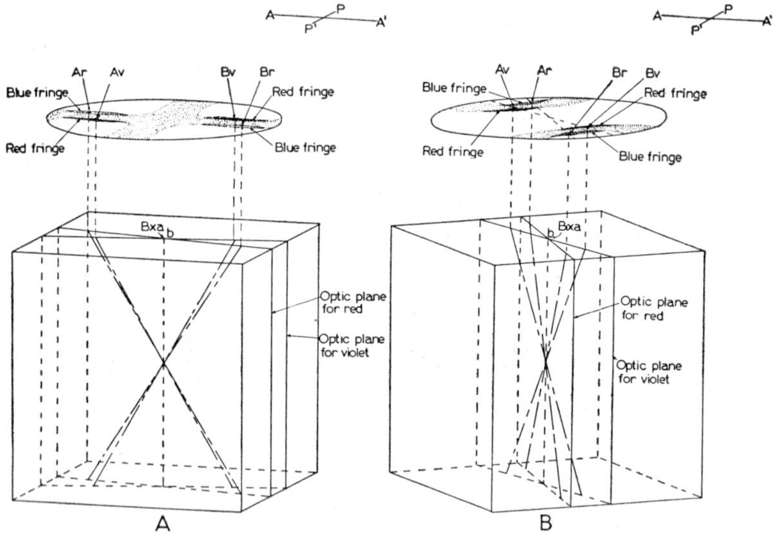

FIG. 5. Crossed dispersion: $r > v$.

A. Parallel position.
B. Forty-five-degree position.

which give acute bisectrix figures in sections cut normal to the b crystallographic axis. The b axis serves as an axis of rotation and the axial planes for the various wave lengths of light assume various positions between the extreme positions occupied by the axial planes for red and violet. As in orthorhombic dispersion, the red color fringes determine the positions of the optic axes for violet.

Crossed dispersion in the parallel position produces similar color fringes on diagonally opposite sides of the isogyres. In Fig. 5, $r > v$ as determined by the fact that the blue fringes are spaced farther apart than the red fringes. In the 45-degree position, Fig. 5*B*, when $r > v$, the red fringes border the convex sides of the isogyres and lie on a diagonal line passing through the center of the field. The blue fringes

display the same relation but lie on the concave sides of the isogyres. When $r < v$, the color fringes are reversed.

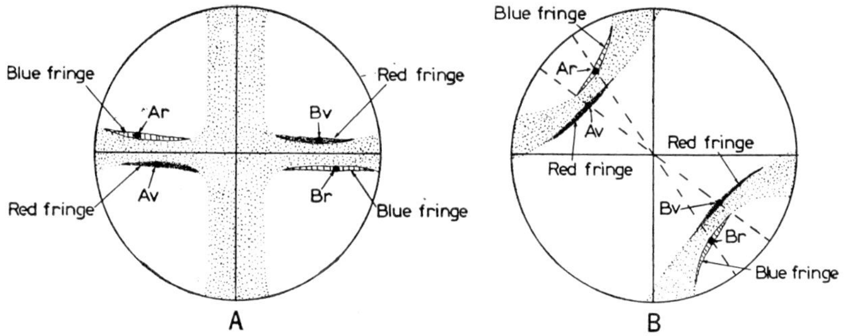

FIG. 6. Crossed dispersion in plan: $r > v$.

A. Parallel position.
B. Forty-five-degree position.

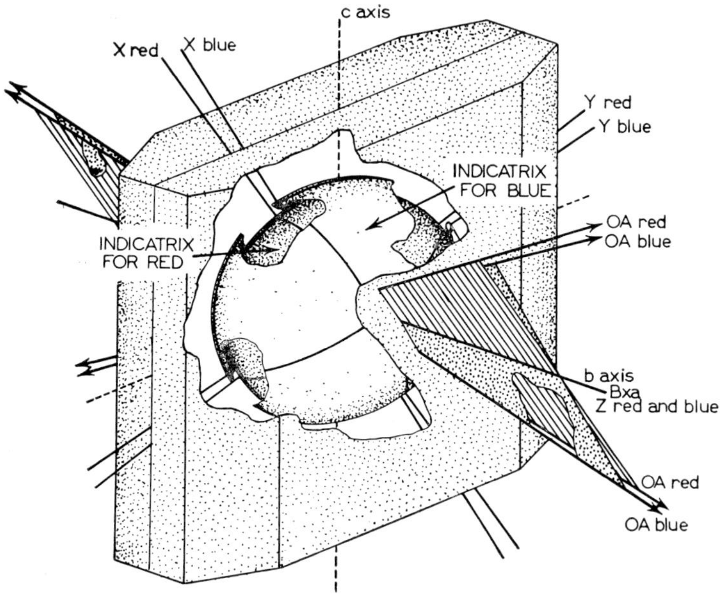

FIG. 7. Diagram of a monoclinic crystal that shows crossed dispersion in an acute bisectrix figure.

Figure 6 shows crossed dispersion in plan. The isochromatic curves are omitted. Figure 7 is a three-dimensional drawing of a monoclinic crystal that would show crossed dispersion in a section normal to the *b* crystallographic axis.

Crossed dispersion, when seen in acute bisectrix figures, assures the observer that he is looking along the b crystallographic axis of a monoclinic crystal.

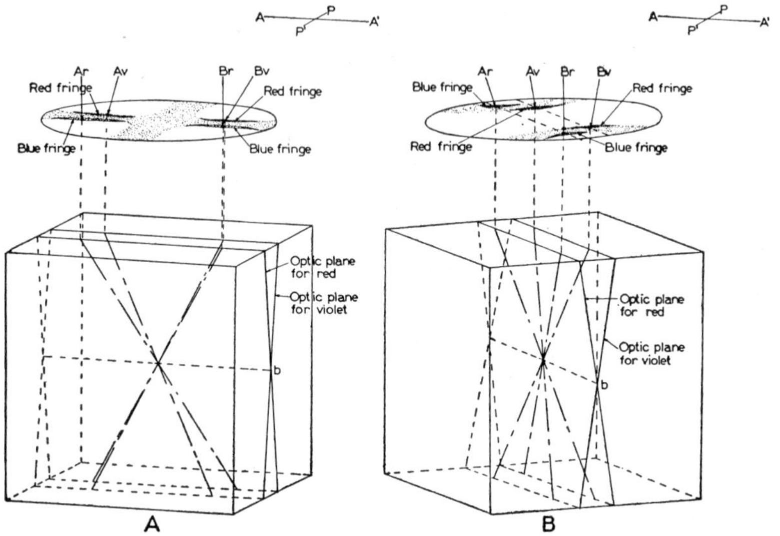

FIG. 8. Horizontal dispersion: $r > v$.

A. Parallel position.
B. Forty-five-degree position.

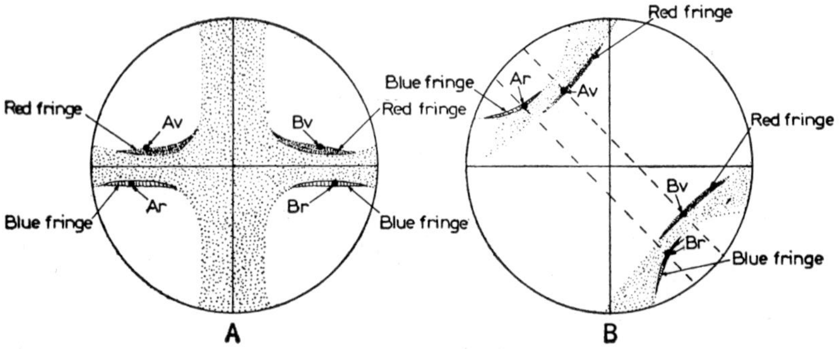

FIG. 9. Horizontal dispersion in plan: $r > v$.

A. Parallel position.
B. Forty-five-degree position.

Figures 8 and 9 illustrate *horizontal dispersion*. This type of dispersion is seen in interference figures of monoclinic crystal sections cut normal to both the acute bisectrix and the plane of crystallographic symmetry. In such sections the obtuse bisectrix parallels the b crys-

tallographic axis. The b axis serves as an axis of rotation, and the optic planes for the various colors assume intermediate positions between the extreme positions of the axial planes of red and violet. In the example shown, $r > v$. Dispersion of the bisectrices is evident

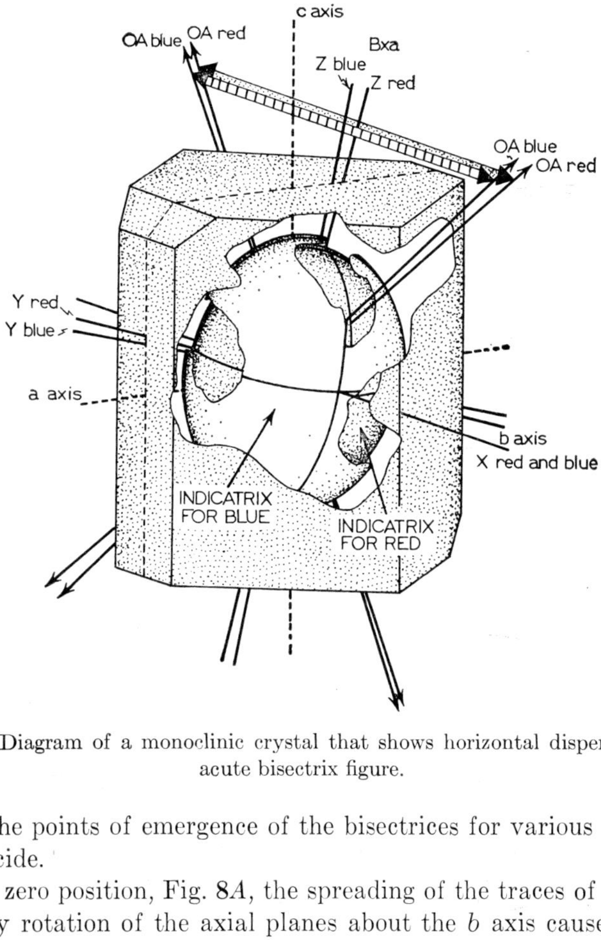

FIG. 10. Diagram of a monoclinic crystal that shows horizontal dispersion in an acute bisectrix figure.

in that the points of emergence of the bisectrices for various colors do not coincide.

In the zero position, Fig. 8*A*, the spreading of the traces of the axial planes by rotation of the axial planes about the b axis causes fringes of the same color to appear in similar symmetrical positions on the isogyres. The color fringes are symmetrical with respect to the symmetry plane of the crystal. When the blue fringes are spaced farther apart than the red fringes, $r > v$.

Observation of horizontal dispersion in an acute bisectrix figure leads to the conclusion that a crystal is monoclinic and that the obtuse bisectrix parallels the b crystallographic axis.

Figure 8*B* shows horizontal dispersion in the 45-degree position.

Figure 9 illustrates horizontal dispersion in plan. A crystal section cut normal to the acute bisectrices of the monoclinic crystal in Fig. 10 shows horizontal dispersion.

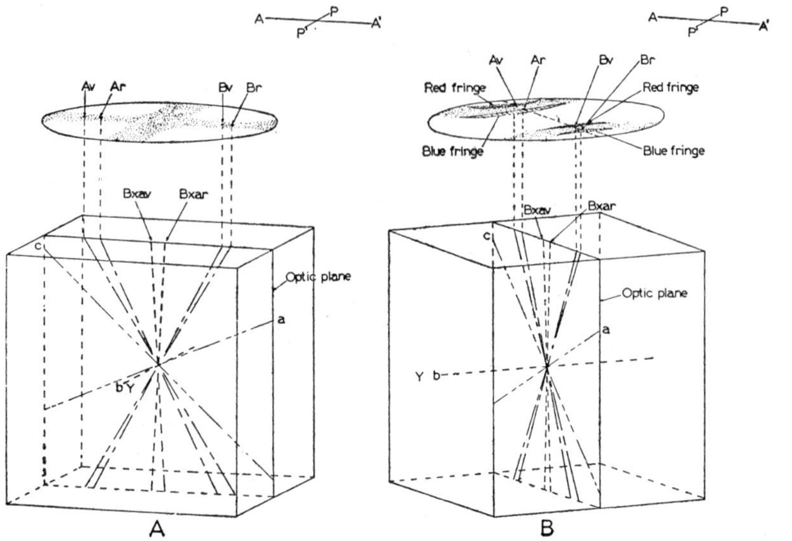

Fig. 11. Inclined dispersion: $r < v$.

A. Parallel position.
B. Forty-five-degree position.

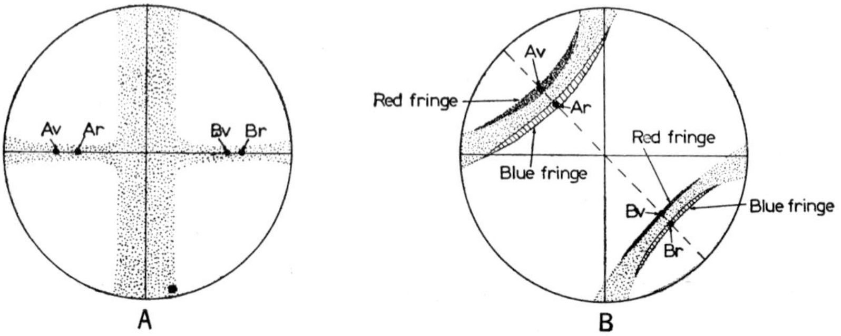

Fig. 12. Inclined dispersion in plan: $r < v$.

A. Parallel position.
B. Forty-five-degree position.

Inclined dispersion produces color effects similar to those shown in Figs. 11 and 12. If the optic normal is parallel to the *b* crystallographic axis, the crystal axis serves as an axis of rotation about which the indicatrices for the various colors rotate until they assume

characteristic positions for a given crystal resulting in dispersion of the bisectrices. The extremes of rotation are the positions assumed by the bisectrices for red and violet. In crystals showing inclined

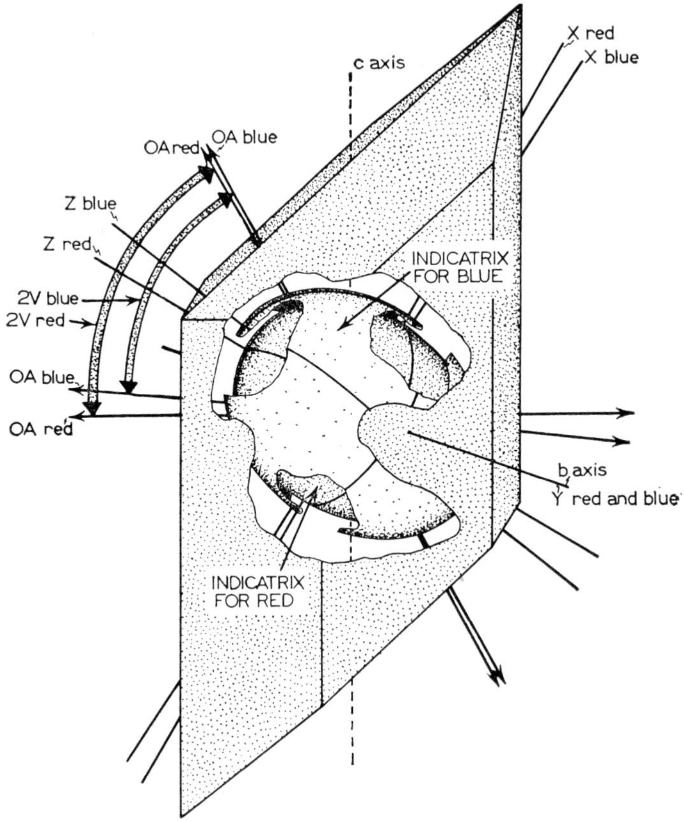

FIG. 13. Diagram of a monoclinic crystal that shows inclined dispersion in an acute bisectrix figure.

dispersion, both the acute and the obtuse bisectrices lie in the crystallographic plane of symmetry.

A section cut from a monoclinic crystal in the parallel position, Fig. 11*A*, gives an acute bisectrix figure for which the isogyre does not differ in appearance from that in a figure showing no dispersive effects. This is explained by the fact that the optic axes for all colors lie in the same plane, the optic plane, and in the zero position no opportunity is afforded for color fringes to appear. However (Fig. 11*B*), color fringes will be seen in the 45-degree position. Many possibilities must be considered, each of which depends on the difference between

the axial angles for red and violet and the extent to which the bisectrices are rotated with respect to each other. Figures 11 and 12 illustrate an example in which $r < v$, and the angular difference between the bisectrices is such that the red fringe appears on the convex side of one isogyre and on the concave side of the other. However, the red fringe, under certain circumstances, might appear on the convex or concave sides of both isogyres. In the latter example, the red and violet fringes on one isogyre would be narrower and spaced closer together than on the other isogyre. If the spacing were the same, the dispersion would be orthorhombic and the crystal would necessarily be orthorhombic. The isogyres and color fringes in inclined dispersion are symmetrical with respect to the optic plane but are not symmetrical with respect to a plane normal thereto. Again this is accounted for by the fact that monoclinic crystals have only one plane of symmetry.

Figure 13 illustrates a monoclinic crystal that would show inclined dispersion in an acute bisectrix figure.

Dispersion in Triclinic Crystals. Triclinic crystals have no planes of symmetry. Accordingly, it is theoretically possible for the indicatrix for each color to assume any position in a crystal, and dispersion of the optic axes, indices, and the bisectrices produces unsymmetrical color fringes in interference figures. Usually the dispersive effects can be described only by assuming a combination of two or more types of monoclinic dispersion. Asymmetrical dispersion, as seen in interference figures, permits classification of a crystalline substance as triclinic.

Probably the best way to visualize dispersion in triclinic crystals is to study the effects in stereographic projection. In general, however, it is sufficient to note the amount of dispersion and to ignore the geometry of the color fringes.

Chapter XVII

MICROSCOPIC EXAMINATION OF NONOPAQUE SUBSTANCES

Introduction. There are many techniques for the microscopic examination of nonopaque substances. The particular technique that is employed depends upon the purpose of the investigation. If it is desired merely to identify a substance by its optical properties, the steps in the procedure are kept to a minimum. If, on the other hand, a precise determination and description of all optical properties is required, the procedure may become very complicated. Optical techniques do not always permit unequivocal identification of crystalline chemical compounds or minerals. One of the chief reasons for this is the fact that many of the optical data appearing in published descriptions and determinative tables are incomplete, and the investigator may have difficulty in locating a described substance having identically the same properties as the material he is studying.

The optical properties of pure, crystalline chemical compounds or minerals and the end members of solid solution series generally are constant and diagnostic. However, many mixed crystal series characterized by limited or complete isomorphism display a partly or completely continuous gradation of optical and other physical properties. In such series, optical data should be plotted on curves showing the optical properties as they are related to composition. In multicomponent isomorphous series, a diagram showing the relationship between chemical composition and optical properties of necessity must be very complex. In order to express adequately the composition, it might be required to assume the presence of three or more end members and to use polyhedral diagrams.

In spite of difficulties that may be encountered in optical studies because of isomorphism or solid solution, the optical technique still remains as one of the most facile methods for identification of nonopaque substances, particularly if the observer has had some experience and has sharpened his judgment as to the effects of isomorphous substitution of one substance for another.

Preferences as to the optical equipment that should be used in measurement of optical constants differ widely. The simplest equipment consists of a petrographic or chemical microscope with the usual

INTRODUCTION   221

lens combinations and accessory plates and a set of standardized
liquid immersion media. For certain purposes a universal stage has
no substitute, but most workers prefer to dispense with complicated
accessory devices and use only the simplest basic equipment. For
accurate work a variable or fixed source of monochromatic light is
required, but in routine identification of substances white light
generally suffices.

In determinative work, preliminary study of physical properties
both in hand specimen and under the microscope serves a useful purpose
and may assist in the identification of a substance which cannot be
distinguished from another substance on the basis of optical properties
alone.

In this text, emphasis has been placed on the immersion technique of
study of nonopaque substances because today this technique is used
more widely than any other. However, the technique may be applied
to the study of crystalline substances in thin sections, if the limitations
of thin-section study are recognized. Students interested in the theory
of universal stage technique are urged to consult the brief summary
in an appendix at the end of this book. The accompanying outline
suggests the important steps in the optical examination of a nonopaque
substance using the immersion technique.

PROCEDURE FOR THE MICROSCOPIC EXAMINATION OF NONOPAQUE SUBSTANCES

I. Preliminary megascopic examination.
   a. Ascertain as many of the following properties as feasible:
      1. Crystallization and crystal habit.
      2. Color.
      3. Luster
      4. Fracture.
      5. Cleavage.
      6. Hardness.
      7. Specific gravity.
      8. Fusibility.

II. Exploratory microscopic examination.
   a. With the upper polarizing prism not inserted determine as many of the
      following properties as feasible in a liquid immersion:
      1. Crystallization and crystal habit.
      2. Color by transmitted light. Selective absorption or pleochroism, if
         present.
      3. Fracture.
      4. Cleavage.
      5. Relief and refractive index relative to immersion medium.
   b. With upper prism not inserted observe whether inclusions or alterations
      are present.

222 MICROSCOPIC EXAMINATION OF NONOPAQUE SUBSTANCES

  c. With upper polarizing prism inserted.
   1. Find out if the substance is isotropic or anisotropic.
   2. Observe twinning if present. If the substance is twinned, note the type of twinning.
   3. Observe the interference color in anisotropic substances and appraise its relationship to thickness, orientation, and birefringence.
 III. If a substance is isotropic, measure the refractive index for white light or one or more specified wave lengths.
 IV. If a substance is anisotropic, make the following observations:
  a. Examine an interference figure to determine whether the substance is uniaxial or biaxial, if this is not already known.
  b. For uniaxial substances determine the following:
   1. Optic sign by any of several available methods.
   2. Sign of elongation or flattening.
   3. Refractive indices for white light or for one or more standard wave lengths.
   4. Birefringence, measured or computed as difference between maximum and minimum refractive indices.
   5. Absorption formula and dichroism.
  c. If a substance is biaxial, determine the following:
   1. Optic sign. If sign is determined from an acute bisectrix interference figure, note the amount and type of dispersion. This procedure helps in determining the crystal system of the substance.
   2. Sign of elongation or flattening.
   3. Refractive indices for white light or for one or more standard wave lengths.
   4. Optic orientation, if crystallographic directions can be identified.
   5. $2V$ and $2E$, either calculated from refractive indices or measured in interference figures.
   6. Pleochroism and absorption formula.
 V. If data obtained in above tests do not suffice for determination, additional chemical or X-ray tests may be needed.

Optical Examination of Crystals or Fragments in Immersion Media.
Particular attention should be paid to the preparation of samples for examination by the immersion technique. Finely divided powders or small, loose crystals may not require preliminary crushing or grinding. However, many coarsely crystalline materials must be reduced to powder before immersion in liquid index media.

The method employed in powdering a sample preparatory to optical examination determines to a considerable extent the ease or difficulty with which the optical constants may be measured. If a substance has no cleavage, simple crushing and grinding with a small steel or agate mortar and pestle is satisfactory. However, if one or more cleavages are present, the powder is produced by crushing or grinding, depending upon what is desired in the subsequent optical measurements. A cleavable substance crushed by a series of sharp blows with the pestle

OPTICAL EXAMINATION OF CRYSTALS 223

yields fragments with flat faces, and, when immersed in a liquid index medium, tends to roll over so as to lie on the flat faces. This result may or may not be desirable. If, on the other hand, the sample is pulverized by a combination of pressure and a gyratory motion of the pestle, the fragments tend to be rounded and are more likely to assume random orientations on the glass slide.

Fibrous and micaceous substances and certain substances with two or more eminent cleavages present a special problem. These substances may be supported in desired positions in immersion media by mixing them with finely powdered cover glass before placing them in the liquid, or they may be subjected to special treatment. Many fibrous and micaceous substances are relatively soft and, if held between two glass slides, may be shaved into thin slices with a razor blade by using the edges of the glass slides as guides. A very small portion of the fibrous or micaceous substance is allowed to extend beyond the edges of the slides, and by successive cuttings several fragments that will lie in the desired position in the liquid immersion may be obtained. Platy or fibrous minerals may be examined by sprinkling fragments on a gel-coated slide before adding the immersion liquid. The gel should not be miscible with the immersion liquid. A solution of water glass generally provides a suitable gel coating on the glass slide.

Many investigators classify the powder by sizes by screening after grinding. A 100-mesh or 60-mesh screen serves this purpose, but for routine work screening of the sample is not generally required. A very small amount of the powder is added to the immersion medium on a glass slide, or the powder is placed on the slide and the immersion liquid is added. A cover glass is placed on top of the immersion for protective purposes. Additional liquid may be introduced under the cover glass by capillarity by touching a drop of liquid to the edge of the cover glass.

To determine all optical constants repeated immersions generally are required. A single crystal may be transferred from one slide to another and rolled into the desired positions by gentle manipulation of the cover glass with the point of a pencil. If a powder is available, it is preferable to use only a few fragments and to roll selected grains so as to obtain desired orientations. If too much powder is added to the index liquid, it will not be easy to observe interference figures because of the difficulty of isolating single grains in the field of the miscroscope.

Optical Examination of Isotropic Substances. Amorphous substances and isotropic crystals have only one refractive index for a

specified wave length. In white light, by using the Becke line and central illumination or oblique illumination in white light, an accuracy of about ± 0.003 can be obtained by the immersion method. The use of monochromatic light permits an accuracy near ± 0.001. For certain types of work it is desirable to ascertain the refractive indices for several standard wave lengths of light. Strained isotropic substances are more or less birefringent.

Optical Examination of Uniaxial Substances. Uniaxial substances belong to the tetragonal or hexagonal systems and possess two principal refractive indices. The indices may be measured in white light by the immersion method with an accuracy near ± 0.003. The index for the ordinary component n_O can be measured in all grains; the index for the extraordinary component n_E can be measured only in grains showing a maximum interference color and yielding a centered flash figure.

The optic sign may be determined when the refractive indices have been measured; or it may be determined in crystals in which the direction of the *c* axis is known or in an optic-axis figure obtained from a grain that remains gray or black during a complete rotation of the microscope stage. In plane-polarized light, with the upper polarizing prism removed, various features such as crystallization, cleavage, etc. may be noted, and once the directions of *O* and *E* have been located, the absorption formula, or dichroic formula, may be noted.

The optical data may be summarized conveniently according to the scheme suggested in the following examples:

	n_{546}	n_{589}	n_{670}	*Dichroism*	
O	2.032 ± .001	2.013 ± .001	1.990 ± .001	Pink	Uniaxial positive; absorption $O > E$, weak
E	2.048 ± .001	2.029 ± .001	2.005 ± .001	Colorless	

	n	
O	1.713 ± .003	Uniaxial negative
E	1.705 ± .003	Colorless

Optical Examination of Biaxial Substances. Biaxial substances crystallize in the orthorhombic, monoclinic, and triclinic systems and have three principal indices of refraction. The maximum and minimum refractive indices, n_X and n_Z, are measured in grains showing a maximum interference color; n_Y may be measured in grains that remain dark between crossed polarizing prisms during a complete rotation of the microscope stage and give centered optic-axis figures, or it may be measured at right angles to the optic plane in crystals giving symmetrical bisectrix figures. In positive crystals giving centered acute

bisectrix figures, n_X may be measured in the trace of the optic plane; in negative crystals, n_Z may be measured in the same manner.

Optic sign may be determined from the refractive indices or from interference figures. The acute bisectrix figure should be examined carefully for dispersion of the optic axes and bisectrices. If dispersion of the bisectrices is absent, the substance is probably orthorhombic. Horizontal, inclined, and crossed dispersion indicate monoclinic crystallization and give valuable clues as to the optic orientation. Triclinic crystals produce asymmetrical triclinic dispersion. The interference figures also may be used to estimate or measure $2V$ and $2E$, or these values may be computed from the refractive indices.

Optic orientation is determined by establishing the relationship between the crystallographic directions and the X, Y, and Z directions of the indicatrix. If it is not possible to identify crystallographic directions by cleavages or crystal outline, complete determination of optic orientation by the immersion technique is very difficult if not impossible.

The absorption formula and pleochroism are noted for the X, Y, and Z directions of the indicatrix at the time that the refractive indices are being measured or during the process of determining the optic orientation.

The optical data may be summarized as indicated in the following examples for orthorhombic and monoclinic crystals:

	n_{Na}	*Orientation*	*Pleochroism*	
X	$1.702 \pm .001$	c	Pink	Orthorhombic Biaxial positive $2V = 84°$ Dispersion $r < v$, weak Absorption $Z > X > Y$
Y	$1.722 \pm .001$	b	Colorless	
Z	$1.750 \pm .001$	a	Rose	

	n	*Orientation*	
X	$1.568 \pm .003$	b	Monoclinic Biaxial positive $2V = 2°$ Dispersion $r < v$, weak Nonpleochroic
Y	$1.569 \pm .003$		
Z	$1.587 \pm .003$	$\wedge c = 21°$	

Triclinic crystals require special treatment. The optic orientation may be shown by means of a stereographic projection which includes the poles of important crystal faces and the poles of important directions of the indicatrix. The optical and morphological crystallographic data may also be given in tables which summarize the angular relationships of all pertinent data in terms of ϕ and ρ angles.

Appendix A

THE UNIVERSAL STAGE METHOD

Introduction. The universal stage method utilizes a multiaxis auxiliary stage which is attached to the rotating stage of the petrographic microscope. Various types have been designed, but only two are in common use: the four- and five-axis stages of German

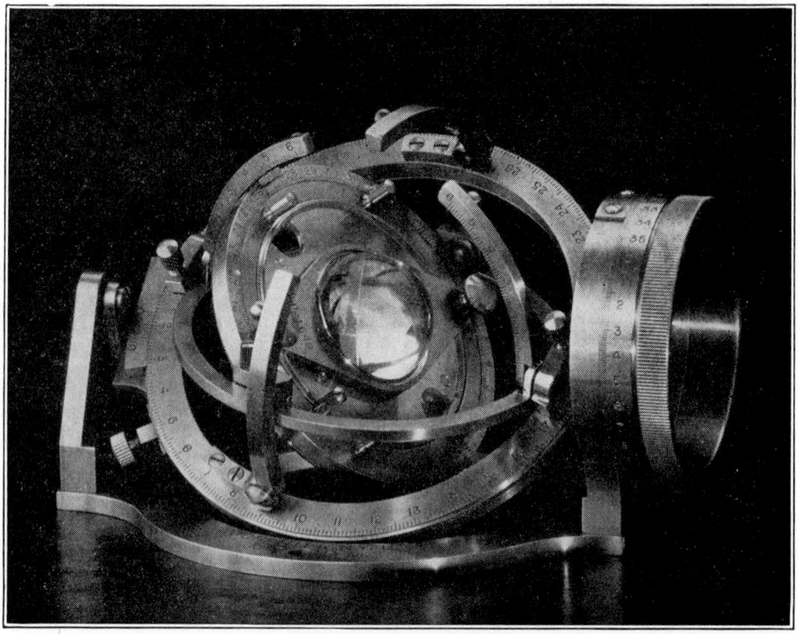

Courtesy Bausch and Lomb Optical Co.

Fig. 1. Five-axis universal stage.

make, and a five-axis stage manufactured in the United States (Figs. 1 and 2). The basic theory of the universal stage technique is simple, but the details of manipulation may be very complex. In essence, the procedure consists of mounting the substance to be examined on a glass slide between two glass hemispheres having a refractive index near that of the substance under scrutiny, and rotating the stage in its various axes so as to bring various optical or crystallographic

planes and directions into critical positions. Although the universal stage finds one of its most important applications in the study of thin sections of rocks, many modifications in design have been made which permit study of crystals or fragments in immersion media.

There are many explicit descriptions of the universal stage technique in the literature, and the student is urged to consult the references listed in Appendix B before attempting serious work. The summary

Courtesy E. Leitz, Inc.

FIG. 2. Four-axis universal stage.

given here is designed to give only an elementary introduction to the theory of the technique: it is a brief discussion that will give the reader a beginner's understanding of the advantages and shortcomings of universal stage study of crystalline materials, and, at the same time, will provide a brief review of certain fundamental concepts of optical crystallography.

Construction of the Universal Stage. Figure 3 indicates diagrammatically the basic elements of construction of the four- and five-axis stages. Both types have an inner glass plate on which the slide mount and the glass hemispheres may be placed so that the crystal plate may be rotated on any or all of four or five axes of the stage. All rotations are measured on conveniently located graduated circles

or arcs of circles. There is considerable difference of opinion as to the proper designation of the axes, but for general purposes the axes may be referred to as east-west, north-south, or vertical. Thus, for the four-axis stage the axes may be designated as follows: inner vertical axis, outer vertical axis, east-west axis, and north-south axis. In the five-axis stage there is an additional east-west axis. The axis of rotation of the microscope stage serves as an additional vertical axis for both types of stages.

In order to eliminate total reflection at the contacts between the hemispheres and the glass plate and slide, oil is applied before putting

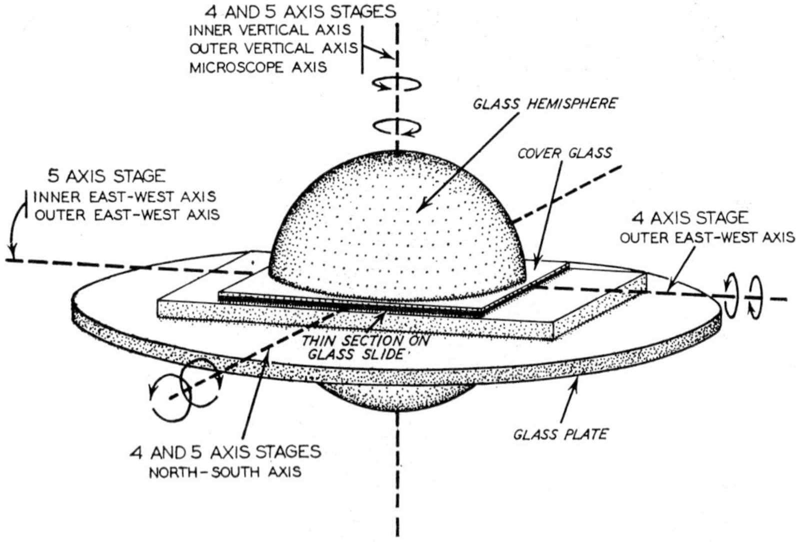

FIG. 3. Thin section mounted between glass hemispheres on universal stage. Diagram indicates rotation axes for four- and five-axis stages.

the hemispheres into position so as to exclude air films that otherwise would be present. All light falls with vertical incidence on the lower hemisphere no matter how the stage is rotated, and if the index of the hemispheres is the same as the crystal plate being examined, light passes through hemispheres and crystal plate without essential deviation. For this reason hemispheres are chosen which have a refractive index as close to that of the crystal plate mounted on the slide as possible. If the refractive indices of hemispheres and crystal plate are different or if the crystal has a high birefringence, there is a certain amount of refraction at the contacts between the crystal plate and the hemispheres, and for very accurate work a correction should

be applied to give the true inclination of the crystal plate with reference to the light passing through it. The methods of determining and applying the correction are described in several of the publications listed in Appendix B.

Plotting of Data. The data obtained from universal stage measurements are plotted conveniently in stereographic projections by using

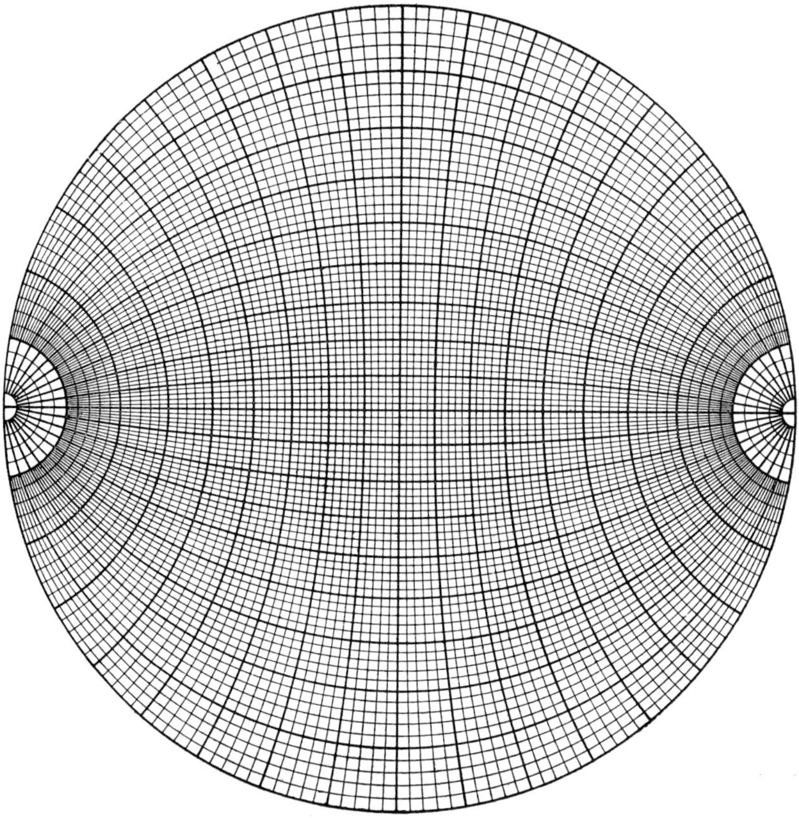

Fig. 4. Wulff stereographic net.

stereographic nets such as the one shown in Fig. 4, a *Wulff net.* The stereographic net is increased in usefulness if appropriately spaced concentric circles and radial lines are added so as to produce a so-called *Fedorow net.* For statistical work a *Schmidt equal-area net* is useful. The Schmidt net is constructed so that a unit area in any position on the net corresponds to a unit area on the spherical projection from which the net is derived.

Data are plotted on the nets in two ways, depending upon the particular type of study that is being made and the preference of the investigator. In some types of work points corresponding to the poles of planes and directions are plotted on the stereographic projection on the assumption that they are derived from points on the upper hemisphere of the spherical projection from which the stereographic projection is derived. In other types of investigations the plotted points correspond to those on the lower hemisphere of the spherical projection. The method of plotting is usually stated in the description of the technique used by an individual observer.

Many types of data may be plotted on the projection. For an individual crystal, poles may be plotted for crystal faces, twin planes, and cleavages, in addition to the points corresponding to the principal planes and directions of the indicatrix. Plotting of points on the stereographic net permits measurement of angles between various directions in the crystal and rotation of projections into any desired position.

Measurement of Uniaxial Crystals. The indicatrix of a uniaxial crystal is an ellipsoid of rotation, and the optic axis is parallel to the *c* crystallographic axis. In general, a crystal in a random position (Fig. 5*A*) shows an interference color. Rotation of the crystal to extinction on a vertical axis (Fig. 5*B*) places a symmetry plane passing through the optic axis of the indicatrix in a position parallel to the vibration direction of either the upper or lower polarizing prism of the microscope. In Fig. 5*B* the symmetry plane includes a north-south axis of the stage, and rotation on an east-west axis in one direction will cause the optic axis to coincide with the axis of the microscope (Fig. 5*C*); rotation in the other direction puts the crystal in a position such that the optic axis parallels a north-south axis.

In routine work with the universal stage it is not possible to observe interference figures because of the low magnification of the objective lens and the relatively large diameter of the glass hemispheres. Accordingly, the optic directions are determined by other methods such as measurement of critical extinction positions. In Fig. 5*C*, for example, it can be determined that the optic axis is parallel to the axis of the microscope because the crystal will remain at extinction during a complete rotation about a vertical axis. The orientation of the crystal in Fig. 5*D* is ascertained by observing a maximum interference color in the crystal when it is rotated to the 45-degree position in a vertical axis.

A crystal in the position shown in Fig. 5*C* permits measurement of n_O; in position 5*D* both n_O and n_E may be measured. The sign can be

determined when it is ascertained whether the variable index for grains in random positions is higher or lower than the fixed index, n_O, or it can be determined with accessory plates by finding out whether the E

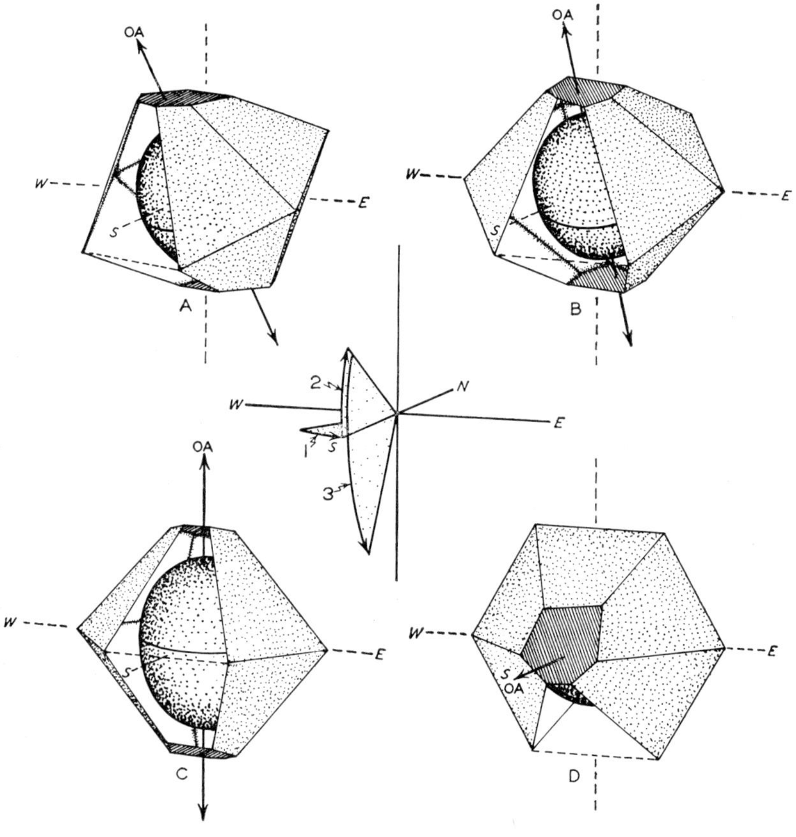

FIG. 5. Rotations of a hexagonal crystal into critical positions.

A. Original random position.
B. Rotated on a vertical axis to the first extinction position.
C. Rotated on an east-west axis so as to align the *c* crystallographic axis with a vertical axis.
D. Rotated on an east-west axis so that *c* axis is parallel to a north-south axis.

component or the *O* component is faster in a crystal oriented as in Fig. 5*D*.

Measurement of Biaxial Crystals. The indicatrix of biaxial crystals is a triaxial ellipsoid. Some idea of the manipulations required to determine optic orientation or measure the optical constants are suggested in a simple example in Fig. 6, which shows an orthorhombic

crystal. Starting in a random position (Fig. 6*A*) the crystal is rotated on the appropriate axis, and angular relationships are noted until enough information is obtained to enable the observer to place the crystal in the various positions indicated in Figs. 6*B*, *C*, and *D*. For

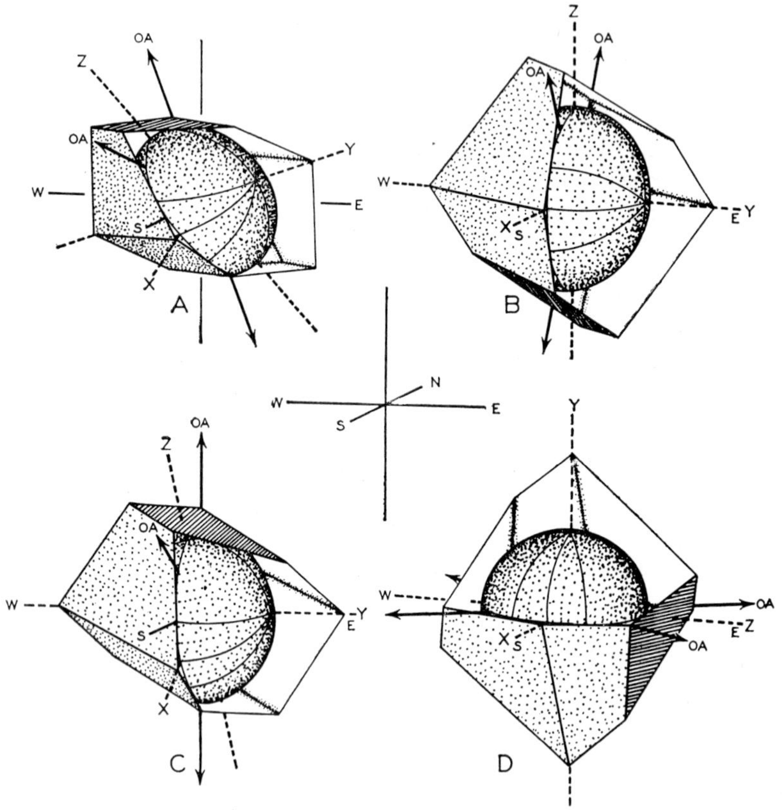

FIG. 6. Rotations of an orthorhombic crystal plate into critical positions.

A. Original random position.
B. Acute bisectrix parallel to a vertical axis; *Y* axis parallel to an east-west axis.
C. Optic axis parallel to a vertical axis.
D. Optic normal (*Y* axis) parallel to a vertical axis.

the four-axis stage the procedure consists essentially of locating the poles of the optical symmetry planes and plotting them on a stereographic projection and then, from the information so obtained, rotating the crystal into the critical positions. The five-axis stage permits critical orientation without going through the plotting procedure.

When the crystal is in the position shown in Fig. 6*B*, the crystal

will remain at extinction upon rotation on either a north-south or an east-west axis, and n_X and n_Y may be measured. Rotation on an east-west axis to the position shown in Fig. 6*D* puts an optic axis parallel to the axis of the microscope. This position is recognized because the grain will remain at extinction during a complete rotation about a vertical axis. Rotation in the opposite direction permits location of the other optic axis and measurement of the optic angle. Measurement of the angle between one optic axis and the acute bisectrix serves the same purpose.

If the crystal is rotated into the position shown in Fig. 6*D*, n_X and n_Z may be measured.

Optic orientation can be determined only if it is possible to locate significant crystal planes or directions by means of crystal faces, cleavages or twinning planes. Sign determination consists essentially of finding out whether the *X* axis or the *Z* axis of the indicatrix is the acute bisectrix.

Appendix B

SELECTED REFERENCES

Theory

Born, M., and Göppert-Meyer, M., "Dynamische Gittertheorie der Kristalle," *Handbuch der Physik,* Bd. XXIV, Zweiter Teil, pp. 623–790, 1933.

Fletcher, L., *The Optical Indicatrix and the Transmission of Light in Crystals,* London, 1892.

Ford, W. E., *Dana's Textbook of Mineralogy,* Fourth Ed., John Wiley & Sons, New York, 1932.

Hartshorne, N. H., and Stuart, A., *Crystals and the Polarizing Microscope,* Arnold, London, 1934.

Niggli, P., *Lehrbuch der Mineralogie,* I, Allgemeine Mineralogie, Zweite Aufl., Borntraeger, Berlin, 1924.

Pockels, F., *Lehrbuch der Kristalloptik,* Teubner, 1906.

Rosenbusch, H., and Wülfing, E. A., *Mikroskopische Physiographie,* Bd. I, 1 Hälfte, Stuttgart, 1924.

Szivessy, G., "Kristalloptik," *Handbuch der Physik,* Bd. XX, pp. 635–900, 1928.

Tunell, G., "The ray-surface, the optical indicatrix, and their interrelation," *Wash. Acad. Sci.,* Vol. 23, pp. 325–338, 1933, and Vol. 28, p. 345, 1938.

Tunell, G., and Morey, G. W., "Some correct and incorrect statements of elementary crystallographic theory and methods in current text-books," *Am. Mineralogist,* Vol. 17, pp. 365–380, 1932.

Tutton, A. E. H., *Crystallography and Practical Crystal Measurement,* Macmillan and Co., London, 1922.

Winchell, A. N., *Elements of Optical Mineralogy,* Part I, Fifth Ed., John Wiley & Sons, New York, 1937.

Wright, F. E., "The transmission of light through transparent inactive crystal plates . . . ," *Amer. Jour. Sci.,* 4th Ser., Vol. 31, pp. 157–211, 1911.

Wright, F. E., "The formation of interference figures . . ." *Jour. Opt. Soc. Amer.,* Vol. 7, pp. 779–817, 1923.

Graphical Solutions and Visual Aids

Chapman, C. A., "A model to illustrate isogyres in interference figures," *Am. Jour. Sci.,* Vol. 238, pp. 805–810, 1940.

Loupekine, I. S., "Graphical derivation of refractive index for the trigonal carbonates," *Am. Mineralogist,* Vol. 32, pp. 502–507, 1947.

Mertie, J. B., Jr., "Nomograms of optic angle formulae," *Am. Mineralogist,* Vol. 27, pp. 538–551, 1942.

Smith, H. T. U., "Simplified graphic method of determining approximate axial angle from refractive indices of biaxial minerals," *Am. Mineralogist,* Vol. 22, pp. 675–681, 1937.

Smith, H. T. U., "Models to aid in visualizing the optical properties of crystals," *Am. Mineralogist,* Vol. 23, p. 629, 1938.

Smith, H. T. U., and Lane, J. H., Jr., "Graphic method of determining optic sign and true axial angle from refractive indices," *Am. Mineralogist,* Vol. 23, pp. 457–460, 1938.

Winchell, H., "A chart for measurement of interference figures," *Am. Mineralogist,* Vol. 31, pp. 43–50, 1946.

Universal Stage Method and Other Special Techniques

Berek, M., "Neue Wege zur Universalmethode," *Neues Jahrbuch,* Beil. Bd., 48, pp. 34–62, 1923.

Berek, M., *Mikroskopische Mineralbestimmung mit Hilfe der Universaldrehtischmethoden,* Borntraeger, Berlin, 1924.

Duparc, L., and Pearce, F., *Traité de Technique Minéralogique et Pétrographique,* Leipzig, 1907.

Emmons, R. C., "The double dispersion method of mineral determination," *Am. Mineralogist,* Vol. 13, pp. 504–515, 1928.

Emmons, R. C., "The double variation method of refractive index determination," *Am. Mineralogist,* Vol. 14, pp. 414–426, 1929.

Emmons, R. C., "Additional comments on the double variation apparatus," *Am. Mineralogist,* Vol. 16, pp. 552–555, 1931.

Emmons, R. C., "The universal stage," Mem. 8, *Geol. Soc. Amer.,* 1943.

Emmons, R. C., and Gates, R. M., "The use of Becke line colors in refractive index determination," *Am. Mineralogist,* Vol. 33, pp. 612–618, 1948.

Fedorov, E. S., "Universalmethode und Feldspatstudien," Part II, *Zeits. Krist.,* Vol. 27, pp. 337–398, 1897.

Fry, W. H., *Petrographic Methods for Soil Laboratories,* U. S. Dept. Agriculture, Tech. Bull. 344, 1933.

Haff, J. C., "Use of the Wulff net in mineral determination with the universal stage," *Am. Mineralogist,* Vol. 25, pp. 689–707, 1940.

Haff, J. C., "Fedorow method of indicatrix orientation," *Colo. Sch. of Mines Quarterly,* Vol. 37, pp. 3–28, 1942.

Knopf, E. B., and Ingerson, E., "Structural petrology," Mem. 6, *Geol. Soc. Amer.,* pp. 226–244, 1938.

Nikitin, W., *Die Fedorow-Methode,* Borntraeger, Berlin, 1936.

Posnjak, E. W., and Merwin, H. R., "The system Fe_2O_3-SO_3-H_2O," *Jour. Am. Chem. Soc.,* Vol. 44, pp. 1965–1994, 1922.

Reinhard, M., *Universal Drehtischmethoden,* Wepf, Basel, 1931.

Slawson, C. B., and Peck, A. B., "The determination of the refractive indices by the immersion method," *Am. Mineralogist,* Vol. 21, pp. 523–528, 1936.

Turner, F. J., "Determination of plagioclase with the four-axis universal stage," *Am. Mineralogist,* Vol. 32, pp. 389–410, 1947.

Tsuboi, S., "A dispersion method of determining the plagioclases in cleavage flakes," *Mineralogical Mag.,* Vol. 20, pp. 108–122, 1923.

Optical Properties of Crystals

Chéneveau, C., "Refractivity of selected solids," *International Critical Tables,* Vol. VII, McGraw-Hill Book Co., New York, 1930.

Iddings, J. P., *Rock Minerals,* Second Ed., John Wiley & Sons, New York, 1911.

Johannsen, A., *Manual of Petrographic Methods,* Second Ed., McGraw-Hill Book Co., New York, 1918.

Johannsen, A., *Essentials for the Microscopical Determination of Rock-Forming Minerals,* Second Ed., Univ. of Chicago Press, 1928.

Larsen, E. S., and Berman, H., "The microscopic determination of the non-opaque minerals," Second Ed., *U. S. Geol. Survey, Bull. 848,* 1934.

Merwin, H. E., "Refractivity of birefringent crystals," *International Critical Tables,* Vol. VII, McGraw-Hill Book Co., New York, 1930.

Rosenbusch, H., and Mügge, O., *Mikroskopische Physiographie der petrographische-wichtigen Mineralien,* Bd. I, Zweite Hälfte, Spezieller Teil, 5 Afl., Stuttgart, 1927.

Rogers, A. F., and Kerr, P. F., *Optical Mineralogy,* McGraw-Hill Book Co., New York, 1942.

Taylor, E. D., "Optical properties in cleavage flakes of rock-forming minerals," *Université Laval,* Géologie et Minéralogie, Contribution 78, 1948.

Tickell, F. G., *The Examination of Fragmental Rocks,* Second Ed., Stanford Univ. Press, 1939.

Winchell, A. N., *The Microscopic Characters of Artificial Inorganic Solid Substances or Artificial Minerals,* Second Ed., John Wiley & Sons, New York, 1931.

Winchell, A. N., *Elements of Optical Mineralogy,* Part II, Third Ed., John Wiley & Sons, New York, 1951.

Wirth, E. H., "Optical crystallographic constants for N. F. VIII," *Nat'l Formulary Committee Bull.,* Vol. XIV, pp. 49–55, 1946.

Index